AS/A-LEVEL YEARS 1 AND 2

STUDENT GUIDE

AQA

Biology

Practical assessment

Jo Ormisher

HODDER
EDUCATION
AN HACHETTE UK COMPANY

The author and publisher would like to thank Richard Fosbery for his assistance with the preparation of this book.

Hodder Education, an Hachette UK company, Blenheim Court, George Street, Banbury, Oxfordshire OX16 5BH

Orders

Bookpoint Ltd, 130 Park Drive, Milton Park, Abingdon, Oxfordshire OX14 4SE

tel: 01235 827827

fax: 01235 400401

e-mail: education@bookpoint.co.uk

Lines are open 9.00 a.m.–5.00 p.m., Monday to Saturday, with a 24-hour message answering service. You can also order through the Hodder Education website: www.hoddereducation.co.uk

This guide has been written specifically to support students preparing for the AQA AS and A-level Biology examinations. The content has been neither approved nor endorsed by AQA and remains the sole responsibility of the author.

Cover photo: Alexander Raths/Fotolia; other photos p. 21 Ian Couchman, Cambridge International Examinations, p. 34 Richard Fosbery, p. 37 Mark Smith, p.40 CDC/Gilda L. Jones, courtesy Public Health Image library, pp.24 and 75 M.I. Walker/Science Photo Library, p.83 blickwinkel/Alamy Stock Photo

Typeset by Integra Software Services Pvt. Ltd, Pondicherry, India

Printed in India

Hachette UK's policy is to use papers that are natural, renewable and recyclable products and made from wood grown in sustainable forests. The logging and manufacturing processes are expected to conform to the environmental regulations of the country of origin.

Contents

Skills Guidance

Questions & Answers

■About this book

This student guide offers advice to help you develop the practical skills required as part of the AQA A-level Biology specification. It covers all the required A-level practicals. Practical skills are assessed in two ways:

- in the AS and A-level examination papers
- in the A-level practical endorsement

Some of the questions in the two AS exam papers and the three A-level exam papers will assess your understanding of the practical skills and your ability to apply them to familiar and unfamiliar contexts. Some of the planning and implementation skills cannot be assessed in a written examination, so will be assessed by your teachers while you are carrying out practicals throughout your course.

During your course you will practise many mathematical skills. This guide covers some of those skills that are needed to analyse and present data in the required practicals.

The guide is divided into two sections:

- **Skills Guidance** begins with a brief guide to some of the general skills and procedures that will be used across the required practical activities. This is followed by detailed guidance on each of the required practicals, including the biology involved, common mistakes and ideas for analysing your results. The required practicals do not have to be carried out in the way described in this book, but these are common examples that allow you to achieve the relevant **Common Practical Assessment Criteria (CPAC)**. They also prepare you for possible exam paper questions on the required practicals. Maths skills are covered within the required practicals where appropriate, although not all of the maths skills from across the whole specification are included in this guide.
- **Questions & Answers** gives examples of some of the types of question that may be set in papers 1, 2 or 3, together with answers written by two students, one of whom makes many errors. There are comments on all the answers. These comments are intended to guide you to write concise answers that show understanding of the key skills in biology practical work.

If you try all the questions in the Questions & Answers section before looking at the answers, you will begin to think for yourself and develop the necessary techniques for answering exam questions and performing well in your practical work, both in the laboratory and in the exam.

Hazard and risk

All practical tasks described in this guide should be risk assessed by a qualified teacher before being performed either as a demonstration or as a class practical. Safety goggles and a laboratory coat or apron must be worn where it is appropriate to do so. The author and the publisher cannot accept responsibility for safety.

■ How the practical element is assessed

The AQA Biology specification requires you to experience teaching and learning opportunities that involve regular practical work.

If you are studying AS biology only then you need to have studied required practicals 1–6 covered in this guide. These practicals will be assessed through the written examinations. This means that you could be asked questions on them in either of the two written exam papers. There is no practical endorsement for the AS qualification, but your teacher may still assess your practical work against the CPAC on page 6.

In the A level course you will complete a minimum of 12 required practical activities. These will then contribute to the practical endorsement element of the A-level biology qualification. The practical endorsement is graded as pass or fail. To pass you need to show evidence of all the required skills (page 6). Your teacher will assess these skills as you complete the practicals. The skills, knowledge and understanding you develop will also be assessed in written examinations in the context of the required practical activities and other practicals. The skills you need to develop can be split into those that can be assessed through written examinations and those that will be assessed by teachers through the practical activities.

Practical skills assessed in written examinations

a Independent thinking:
   ■ Solve problems set in practical contexts.
   ■ Apply scientific knowledge to practical contexts.
b Use and application of scientific methods and practices:
   ■ Comment on experimental design and evaluate scientific methods.
   ■ Present data in appropriate ways.
   ■ Evaluate results and draw conclusions with reference to measurement uncertainties and errors.
   ■ Identify variables, including those that must be controlled.
c Numeracy and the application of mathematical concepts in a practical context:
   ■ Plot and interpret graphs.
   ■ Process and analyse data using appropriate mathematical skills. These skills are described in Section 6 of the AQA specification.
   ■ Consider margins of error, accuracy and precision of data.
d Instruments and equipment:
   ■ Know and understand how to use a wide range of experimental and practical instruments, equipment and techniques appropriate to the knowledge and understanding included in the specification

Use of apparatus and techniques

As you complete the required practicals, you are expected to develop your skills in the use of a wide range of apparatus and techniques (ATs). The ATs listed below are compulsory and should all have been completed for the full A-level course.

How the practical element is assessed

ATa Use appropriate apparatus to record a range of quantitative measurements (to include mass, time, volume, temperature, length and pH).

ATb Use appropriate instrumentation to record quantitative measurements, such as a colorimeter or potometer.

ATc Use laboratory glassware apparatus for a variety of experimental techniques, to include serial dilutions.

ATd Use a light microscope at high power and low power, including use of a graticule.

ATe Produce scientific drawings from observation with annotations.

ATf Use qualitative reagents to identify biological molecules.

ATg Separate biological compounds using thin-layer/paper chromatography or electrophoresis.

ATh Use organisms safely and ethically to measure:
- plant or animal responses
- physiological functions

ATi Use microbiological aseptic techniques, including the use of agar plates and broth.

ATj Use instruments safely to dissect an animal organ or a plant organ.

ATk Use sampling techniques in fieldwork.

ATl Use ICT such as computer modelling or a data logger to collect data, or use software to process data.

The practical skills assessed by your teacher

Your teacher will assess you against the **Common Practical Assessment Criteria (CPAC)**. These define the minimum standard required to achieve a pass.

You will need to keep an appropriate record of your practical work, including your assessed practical activities. This can be done using a lab book provided by your teacher. Make sure that all of your written work is clearly presented, as it will form part of your teacher's assessment of the CPAC. You will also use it when revising for your final exams.

If you demonstrate the required standard across all the requirements of the CPAC you will receive a 'pass' grade. Your teacher will make a judgement of your practical competence using the CPAC and will need to include comments to justify the decision made. To achieve the pass grade you need to consistently and routinely exhibit the competencies listed in the CPAC before the completion of the A-level course. A major component of the practical endorsement is the ability to apply your skills to the practical work independently, which means that you cannot just rely on your teacher telling you what to do.

The **CPAC** are:

1 Follows written procedures.
2 Applies investigative approaches and methods when using instruments and equipment.
3 Uses a range of practical equipment and materials safely.
4 Makes and records observations.
5 Researches, references and reports.

The AQA A-level Biology specification gives further details of the CPAC requirements.

Skills Guidance

■ Practical advice

Following written procedures

When carrying out practical work you usually follow a set of written instructions. Before you start the practical work read these through to the end. Annotate the instructions to help you understand them. Then check that you have all the apparatus and materials listed. When you are ready to start, read the first instruction carefully and carry it out. Put a tick by each instruction when you have completed it. Proceed carefully through the rest of the instructions, double-checking that you are sticking to them. This is important to ensure the collection of accurate data and to make the practical activity safe. Following instructions shows competency in CPAC 1.

Investigative approaches

You may be asked to plan an aspect of an investigation in the exam papers and you will probably plan and carry out a complete investigation as part of your practical work so that you can be assessed on your investigative approach for CPAC 2.

Throughout your course and in the examinations you will be tested on the skills involved in planning, such as identifying variables, stating a hypothesis, writing a method or explaining how to collect results and/or analyse them. Writing full plans during your course will prepare you well for these questions. Here are some steps to follow while planning an investigation:

1 **Identify a question to answer and write a hypothesis**. Read the information provided and look for clues. Write a question that you have to answer by experiment and then write a hypothesis, which is a clear statement about what you think will happen. You must write a **testable hypothesis** — one that you can test by experiment.

2 **Carry out some research**. Use sources of information to read about the problem you are trying to solve. Look for ideas to help your planning and decide how results will be analysed statistically.

3 **Write a null hypothesis**. If you are going to use a statistical test to analyse your results then you must rewrite your hypothesis as a negative statement, known as a **null hypothesis**. This states the opposite of your hypothesis. Your experiment must test the idea that there will be *no effect*.

4 **Identify the independent, dependent and control variables for the investigation:**
   - **Independent variable** (IV) — the variable you change.
   - **Dependent variable** (DV) — the variable you measure or observe.
   - **Control variables** (CV) — the variables that are kept constant, because they might affect the values of the DV.

Skills Guidance

5 **Decide on a strategy for your experiment**. This is a brief outline of a method that tests your hypothesis.

6 **Choose apparatus and materials that are appropriate**. You should be able to select appropriate equipment to collect accurate data. Often it is a good idea to justify your choice of the main items of apparatus. An important consideration is the resolution of the apparatus you use.

Resolution is the smallest change in quantity that can be measured with the apparatus that you are using. This refers both to the apparatus used to measure out quantities as part of the procedure and the apparatus used to collect results. It refers to the number of significant figures (or decimal places) in readings.

When measuring out volumes, for example, you might use a syringe, pipette, burette or measuring cylinder. A burette measures to the nearest $0.1\,cm^3$ so you would select a burette if you were measuring $0.5\,cm^3$ of a solution, but this resolution would not be necessary if you wanted to measure $100\,cm^3$.

When collecting results, a balance might weigh to the nearest $0.1\,g$; some balances are more sensitive and weigh to the nearest $0.01\,g$. If measuring change in length you would use a ruler that measures to the nearest millimetre. These are measures of resolution in results taking. Recording to the nearest $0.01\,g$ gives a higher resolution than measuring to $0.1\,g$. Similarly, measuring to the nearest millimetre gives a higher resolution than measuring to the nearest centimetre. In this context resolution means the number of significant figures or decimal places to which values are expressed.

Stop clocks and bench timers can often measure to a hundredth of a second ($0.01\,s$). It is highly unlikely that you could time a colour change or other event to this degree of resolution, so it is better to express such results to the nearest second.

7 **Expand the strategy into a detailed procedure, using numbered steps**. Avoid using continuous prose — it is easier to follow a series of instructions. Notice that numbered points allow you to include instructions such as 'repeat step 6'. The procedure must describe how results are to be collected.

8 **Explain how results are to be presented**. A good way to do this is to draw a table with clear column headings, including units.

9 **Explain how results are to be analysed**. This includes:
- data processing
- graphs
- statistical tests

10 **Plan and carry out a preliminary investigation** to trial your ideas. You may have to modify your procedure as a result. If so, record the modifications you made and why you made them in your report.

Practical tip

If you choose to use a colorimeter, you could say that this gives quantitative results that are easier for others to reproduce than if you use colour standards (charts showing expected colours). You will use a colorimeter in required practical 11.

Practical tip

Notice that resolution also applies to microscopy. It refers to the smallest distance that can be detected when using a microscope.

Safety

Planning and carrying out practical work safely is essential and demonstrates competency in CPAC 3. If you work unsafely, you are not only putting yourself at risk, but also other people in the class.

A risk assessment should be completed before any practical activity is undertaken. The responsibility for this lies with your school or college, but you may be asked to write a risk assessment for one or more of the required practicals. Your school or college should have documents on safety published by an organisation known as CLEAPSS (Consortium of Local Education Authorities for the Provision of Science Services) that you could refer to for details of possible risks and how to control them.

The following is some general safety advice, but if you are unsure of something, it is important that you check with your teacher.

- Keep your work area well organised and tidy.
- Use one area for wet work and keep another area dry for writing your notes. Do your practical work over the bench, not over your papers.
- Wear protective goggles/spectacles when using liquids or when cutting anything.
- Make sure that you are familiar with hazard warning symbols and know how to respond to them.
- Take special care when using knives, scalpels, glassware, chemicals, Bunsen flames, hot water etc.
- Inform the teacher or technician immediately if you have an accident.
- Clear up any spillages or broken glass immediately.

Table 1 shows examples of poor laboratory practice, together with equivalent good practice.

Practical tip

Your teacher will assess your approach to working methodically and safely while you carry out your practical work.

Practical tip

Syringes should be washed out with water and then with a small volume of the liquid that you are going to dispense.

Table 1 Examples of poor and good laboratory practice

Poor practice	Good practice
Using incorrect apparatus or equipment without realising*, for example using a $10\,cm^3$ syringe instead of a $1\,cm^3$ to dispense $0.5\,cm^3$ of a liquid	Choosing the correct apparatus/equipment/chemicals from those provided so that the correct results are obtained
Using the same syringe to dispense different liquids without realising* so that contamination occurs	Using separate syringes when necessary and keeping syringes separate once used (for correct re-use) or washing out a syringe between using to dispense different solutions
Cutting slices or cubes or sections carelessly so that sections are of uneven thickness or cubes are of unequal size	Using cutting equipment (e.g. scalpels and knives) and measuring equipment (e.g. rulers) with care to produce slices and cubes of correct sizes
Allowing fluids to drip off the outside of beakers/tubing/stirring rods/tissue samples into other solutions so that there is the risk of cross-contamination	Rinsing and drying equipment when necessary, clearing off spills on the outside of beakers; keeping different items in clearly defined areas on the bench
Haphazard use of the stop clock/bench timer so that incorrect times are recorded; samples are not taken at correct time intervals	Checking how to use the stop clock/bench timer before starting; careful checking of times; re-setting back to zero when required
Filtering suspensions through a filter funnel where the filter paper has not been folded correctly/has a tear/has not been fitted into the funnel correctly	Folding a piece of filter paper and then opening it out into the filter funnel; filtering suspensions so that a clear solution, the filtrate, runs through and the precipitate/larger insoluble particles remain on the filter paper

* Realising a mistake and asking for fresh syringes is considered good practice.

Recording observations

Making accurate observations and recording them in a suitable form is an important skill that provides evidence for CPAC 4. In most cases, results and observations will be recorded in tables.

Before you start to draw a table, decide what you need to record. Decide on how many columns and how many rows you will need. Make sure you have read all the practical instructions before you draw the table outline. Follow these rules:

- The first column should be the independent variable, the second and subsequent columns should contain the dependent variables.
- Write clear and detailed headings for each column.
- The headings of the columns must include the relevant units. There should be no units in the body of the table.

Practical tip

Units are separated from the description of the variable by a forward slash (/). The slash should *not* be used to mean 'per' in compound units. For example, do *not* write g per cm^3 as g/cm^3, but as g cm^{-3}.

- Data should be ordered so that trends and patterns can be seen — it is best to arrange the values of the independent variable in ascending order.
- Use plenty of space — do not make the table too small.
- Leave some space to the right of the table in case you decide you need to add more columns.
- Use a pencil and ruler to draw lines between the columns and between the rows. Rule lines around the whole table.

Evaluating procedures

A sound understanding of the required practicals is essential, so a thorough analysis of your results will prepare you for the exams as well as providing evidence for CPAC 5. When evaluating an experimental procedure it is important to consider the way in which the procedure was carried out and the quality of the data collected. You need to ask yourself the question: 'can I have confidence in my data?' If you do not have confidence in the data then you cannot have confidence in the conclusion(s) that you make.

The first thing to do when evaluating is to consider the procedure that you followed. Is it possible that there were any **measurement errors** in the method? There are two types of error:

- **Systematic errors** are always the same throughout the investigation. A common type of systematic error is that the measuring device may give readings that are incorrect by a certain value. It could be that one of the control variables is always incorrect by the same quantity. If there are small systematic errors (that are always the same) then the data may be precise, but not accurate. The effect of these errors is to overestimate or underestimate the true values of the dependent variable.

Practical tip

There are plenty of examples in this guide to help you become proficient in drawing and completing tables of results.

Practical tip

The general rules for drawing tables apply whether producing a hand-drawn table in your lab book, or using a spreadsheet program.

Practical tip

Never jot your results down on pieces of rough paper. Draw a table to record your raw data in your lab book so that you can refer back to it if necessary.

Practical tip

The words in **bold** type are important key terms that you must understand and use. Look for the use of these terms in the Questions & Answers section.

■ **Random errors** occur when you do not carry out the procedure in exactly the same way each time. You may also read the apparatus in a slightly different way each time you take a reading. These errors affect some of the results, but not all of them. They do not always affect the results in the same way. Random errors could be the result of the variation in biological material.

Do not think of errors as mistakes. Even in a perfectly conducted investigation, there will be errors.

You should consider the control variables involved in the investigation. These should be kept constant, or monitored if it is not possible to keep them constant. If these variables are not controlled then they may influence the results; they are called **confounding variables** or **uncontrolled variables**. Sometimes, particularly in fieldwork studies when you cannot control certain abiotic factors, you may be aware of such variables and then 'take them into account' when analysing and interpreting results.

Control variables should not be confused with a **control experiment**. A control experiment is important to show that your results are valid. For example, if you are investigating the effect of temperature on the hydrolysis of starch by amylase, you need to show that it is the amylase that is causing the hydrolysis rather than the buffer solution or heat. A control experiment may involve using boiled amylase to show that hydrolysis is due to enzyme action.

There are several terms with specific meanings that are used when discussing the quality of the procedure and the results obtained:

■ **Accuracy** is a measure of how close a result is to an accepted 'true' value. In biological investigations the true value is not always known. In some cases results can be checked with sources of data. For example, tidal volume readings should be about 500 cm^3, the water potential of the blood should be equivalent to 0.9% sodium chloride solution (−3.86 MPa), but the water potential of plant tissues varies considerably and there is no specific value against which results can be checked.

■ **Anomalous results** are results that do not fit the trend. They can be:
  ■ replicate results that differ significantly from others
  ■ results for one value of the independent variable that do not fit the overall trend

Accuracy refers to how close a measurement is to its true value.

Practical tip

Take special care not to overuse the term accuracy. There are very few biology investigations when you can say what the true value(s) should be.

Precision refers to the spread of measurements about the mean value.

Practical tip

Anomalous results should not be ignored. You should initially repeat the result and see if you get a similar reading. If you do, you should then consider possible causes; for example:
■ Is the solution from the same batch?
■ Is it due to experimental error?
■ Have you used a different piece of measuring equipment?

Practical tip

Calculating a **running mean** is a good way to check that you have enough replicate results. Calculate the mean after you have collected each replicate and continue doing this until it remains near constant.

■ **Precision** is a measure of the closeness of agreement between individual results obtained using the same procedure under exactly the same conditions. However, closeness of replicates does not mean that the data are close to the true value.

■ **Repeatable results** are replicate results that are in close agreement. You can use mathematical methods to help evaluate the variation in replicate results. In A-level practical tasks it is usual to carry out three repeats or replicates for each value of the independent variable if time and materials permit. These should be carried out separately from each other, using exactly the same experimental procedures.

- Results are **reproducible** if someone else who repeats the investigation obtains the same results as you. You can only comment on this in response to a question if you are given results from another person.

- **Uncertainty** is half the smallest graduation on the apparatus. For example, if the smallest division on a syringe is $1.0\,cm^3$ then the uncertainty would be $\pm0.5\,cm^3$. If you are certain that you have started measuring at 0, then the uncertainty applies where you take the measurement, so $6.3\,cm^3$ is expressed as $6.3 \pm 0.5\,cm^3$. If you are not sure that you started measuring exactly at 0 or you have started at a measurement other than 0 (for example when using a burette) the uncertainty applies at both ends, so it is multiplied by two as there is an error at each end, giving an uncertainty of $\pm1.0\,cm^3$.

  It is possible to calculate the **percentage error** for the apparatus you used for measuring your results. If you have collected $5.0\,cm^3$ of a gas and measured the volume with a syringe that has graduations every $1\,cm^3$, your uncertainty is $\pm0.5\,cm^3$. This makes the percentage error:

$$\text{percentage error} = \frac{0.5}{5.0} \times 100 = 10\%$$

- **Validity** refers to both individual measurements and to the whole procedure. If you have a **valid result**, then you know that you measured what you set out to measure. If you have a **valid investigation** then you have measured what you intended to measure and you can be confident that changing the independent variable leads to changes in the dependent variable.

- When you make a conclusion about an investigation then you can make a judgement about the extent to which the evidence collected supports that conclusion. In doing so you are expressing **confidence** in your conclusion. If asked to comment on the confidence in a conclusion then you should consider the following:
  - the limitations in the procedure
  - any uncontrolled variables
  - the effects of errors (systematic and random) on the results
  - the repeatability of the results
  - the precision of the data collected
  - the accuracy of the results

Give some positive aspects of the investigation first, followed by some criticisms. You should refer to specific aspects of the procedure and results, rather than using vague comments such as 'my conclusion is valid because my results are precise and accurate' — this is meaningless without supporting information. For example, you can say that your results are precise because the replicates are close together and there are no anomalous data. Your results may be accurate because they all agree with an expected trend.

Always quote some examples of your raw or processed data in support of your arguments. You can also comment on the resolution of your apparatus, for example by saying that you have measured the dependent variable to a high degree of resolution (e.g. weighing to $0.01\,g$). Resolution refers to the smallest change in the quantity being measured that an instrument can detect (page 8). If you measure $0.2\,g$ of a chemical on a balance that measures to the nearest $0.1\,g$, the balance will not detect a change in mass to $0.19\,g$ or $0.23\,g$ due to the resolution of the equipment.

■ Required practical activities

Required practical 1

Factors affecting the rate of enzyme-controlled reactions

Background information

This practical requires you to investigate the effect of one variable on the rate of an enzyme-controlled reaction. You will have studied the induced-fit model of enzyme action and the properties of enzymes relating to their tertiary structure. The factors that affect the rate of an enzyme-controlled reaction are:

- enzyme concentration
- substrate concentration
- concentration of competitive and non-competitive inhibitors
- pH
- temperature

The investigation that you carry out will depend on the apparatus and chemicals available, but the most common variables to investigate are pH and temperature.

The enzymes below are popular choices in schools and colleges:

- **Amylase** — hydrolyses starch to maltose and the end-point of the reaction can be clearly visualised using iodine in potassium iodide solution. The end-point is when the blue-black colour no longer appears.
- **Catalase** — catalyses the breakdown of hydrogen peroxide into water and oxygen. The oxygen can be collected and the rate of reaction determined by measuring the volume of gas produced in a specific period of time or the time taken to collect a specific volume of gas. Figure 1 shows some of the different methods that can be used to collect gas.
- **Protease**, for example trypsin — hydrolyses insoluble protein to small, soluble peptides. The time taken for a cloudy protein solution to clear can be measured.

Guidance through the practical

As with any investigation, it is essential to identify the independent, dependent and control variables.

The **independent variable** will be one of the five factors that affect the rate of an enzyme-controlled reaction. You may be provided with specific values to use — for example, you may be told to use pH 4, 5, 6, 7 and 8 — or you may have to decide suitable values for yourself based on your knowledge of enzyme action.

The **dependent variable** will be a measurement of the time taken for the reaction to complete. This will either be the time taken for the **substrate** to be used up, or the time taken for the **product** to form.

The **control variables** are the ones that need to remain constant. You should be able to state how and why you controlled these variables because identification of control variables is one of the practical skills that will be assessed in the written papers. If your independent variable is temperature, you need to control all of the other factors that could affect the rate of reaction, for example pH using a buffer solution.

> **Exam tip**
>
> Always refer to 'iodine in potassium iodide solution' or 'iodine solution', never just 'iodine', because iodine is a solid.

> **Exam tip**
>
> A question on practical work may include the use of reagents to identify biological molecules. Biuret reagent is used to test for protein. If no protein is present, the solution remains blue. If protein is present, the solution turns purple.

> **Exam tip**
>
> Always refer to 'volume' rather than 'amount' of oxygen produced. The word 'amount' is ambiguous and is often wrongly used when volume, concentration or number is the correct term to use.

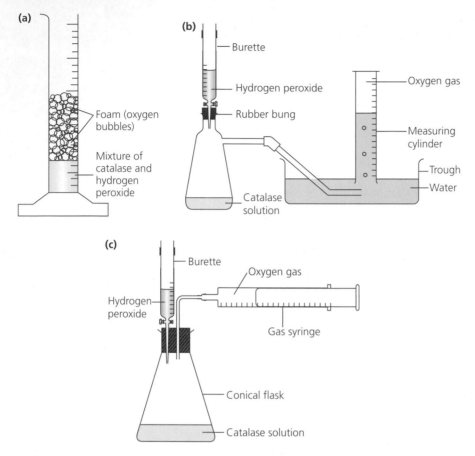

Figure 1 Different techniques for measuring oxygen production with catalase

The written papers may assess your ability to consider margins of error, accuracy and precision of data. All measurements that you make have some **uncertainty** so measurements are often written with the uncertainty. For example, when temperature is measured the scale usually has intervals of 1°C, so the uncertainty is half of this interval (0.5°C) and so a measurement would be written as, for example, 20.0 ± 0.5°C.

You should consider the uncertainty of the measurements you need to make when selecting suitable apparatus and instrumentation (AT a and AT b). If you need to measure 1 cm³ of solution, a piece of equipment with intervals of 0.1 cm³ will give a smaller uncertainty (±0.05 cm³) than a piece of equipment with intervals of 0.5 cm³ (±0.25 cm³).

Presentation of results from the practical

Before starting the practical, you should draw a suitable table in which to record your measurements. When drawing a results table, follow the rules given in the practical advice section (page 10). Figure 2 shows an example of a student's results table from an investigation in which the student compared the activity of amylase from two sources.

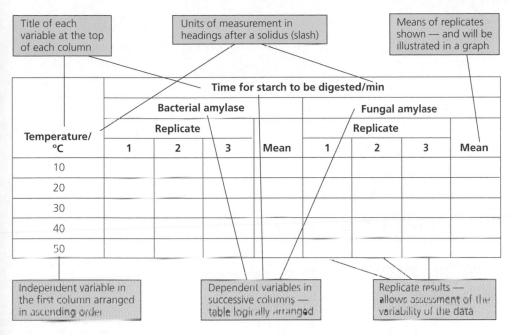

The following labels surround the table in Figure 2:

- Title of each variable at the top of each column
- Units of measurement in headings after a solidus (slash)
- Means of replicates shown — and will be illustrated in a graph
- Independent variable in the first column arranged in ascending order
- Dependent variables in successive columns — table logically arranged
- Replicate results — allows assessment of the variability of the data

Temperature/ °C	Time for starch to be digested/min							
	Bacterial amylase				Fungal amylase			
	Replicate				Replicate			
	1	2	3	Mean	1	2	3	Mean
10								
20								
30								
40								
50								

Figure 2 A table showing the effect of temperature on starch digestion, with guidance on table structure

The data you collect for the dependent variable will be a measurement of the time taken for the reaction to reach completion. The investigation requires you to determine the **rate** of the reaction, so you will need to process these data. The rate is the speed of the reaction, so it will be per unit time, for example per second (s^{-1}) or per minute (min^{-1}). If you have found the volume of oxygen produced in a specified time, you would determine the rate using volume/time. You need to include suitable units for rate, so if you collected $10\,cm^3$ of oxygen in 5 minutes, the rate of reaction would be $2\,cm^3$ per minute, expressed as $cm^3\,min^{-1}$.

In other investigations, you may just have a measurement of time. To express this as a rate, you need to calculate the reciprocal of the time, which is 1/time. Again, you need to include the unit — s^{-1}, min^{-1} or h^{-1}, depending on the measurements taken.

Once you have processed your data, you should present them as a graph. If you have investigated a continuous variable (temperature, pH, enzyme concentration or substrate concentration) your results should be presented as a line graph. If you have investigated the effect of an inhibitor — for example, rate of reaction *with* an inhibitor and rate of reaction *without* an inhibitor — then a bar chart would be appropriate. Your ability to present data graphically will be assessed by examiners in the written papers and also by your teacher as part of the practical endorsement (CPAC 4).

The following rules should be followed when plotting your graph:

- The independent variable (the factor changed) is on the *x*-axis. The dependent variable (the rate of reaction) is on the *y*-axis.
- Both axes are fully labelled and have correct units.
- An appropriate scale has been chosen for each axis.
- Points have been plotted accurately using saltire crosses (×) so they can be seen clearly.
- Each point has been joined with a straight line.

Exam tip

Use the term **mean** rather than average. Average refers to a central value, which could be the mean, median or mode.

Exam tip

When calculating the reciprocal, you could also use 100/time or even 1000/time so that you do not get a rate with too many decimal places.

Exam tip

Always choose a scale that is easy to work with, for example multiples of 2, 5 or 10. Avoid using scales that are difficult to work with, such as intervals of 3.

Skills Guidance

Figure 3 shows a graph that has been plotted following these five rules, whereas the graph shown in Figure 4 does not have a suitable x-axis scale and the line has been extrapolated.

Exam tip

Always join points with a straight ruled line if you cannot be certain of the intermediate values.

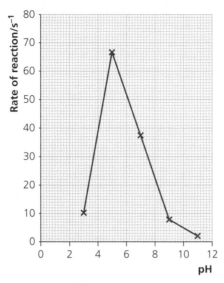

Figure 3 A graph to show the effect of pH on the activity of amylase. The graph has the axes the correct way round, a suitable scale, clearly labelled axes and correctly plotted points joined with straight lines

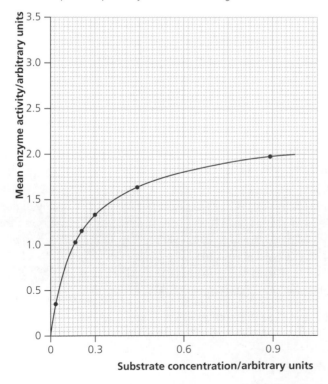

Figure 4 A graph to show the effect of substrate concentration on urease activity. The graph has a poor choice of x-axis scale that is non-linear (does not increase in equal increments)

Analysis/evaluation of the results

Once you have plotted your graph, you need to use your knowledge of enzyme action to explain your results. You should describe the trend or overall pattern of your results using terms such as increase/decrease, plateau/level off, maximum and directly proportional.

Figure 5 shows some of the common trends seen in graphs, together with descriptions of the trends.

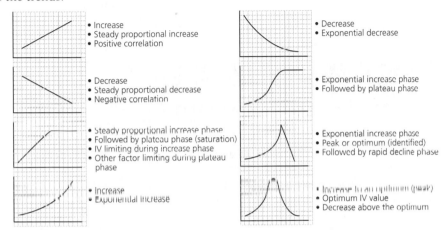

• Increase
• Steady proportional increase
• Positive correlation

• Decrease
• Exponential decrease

• Decrease
• Steady proportional decrease
• Negative correlation

• Exponential increase phase
• Followed by plateau phase

• Steady proportional increase phase
• Followed by plateau phase (saturation)
• IV limiting during increase phase
• Other factor limiting during plateau phase

• Exponential increase phase
• Peak or optimum (identified)
• Followed by rapid decline phase

• Increase
• Exponential increase

• Increase to an optimum (peak)
• Optimum IV value
• Decrease above the optimum

Figure 5 Graphs showing some common trends and descriptive terms

When you are explaining the shape of a graph, you may find it helpful to number the main sections of your graph and then describe each section.

The graph shown in Figure 6 has three main sections, labelled A, B and C.

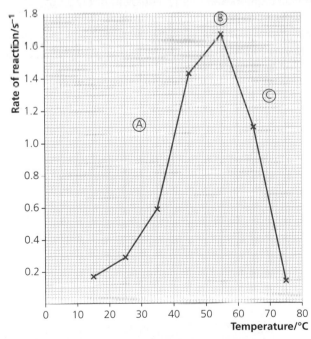

Figure 6 Graph to show the effect of temperature on the rate of an enzyme-controlled reaction

Section A of the graph shows the rate of reaction increasing between 15°C and 55°C. You should explain this by referring to the increased temperature and its effect on the kinetic energy of the molecules, the number of successful collisions and the formation of enzyme–substrate complexes.

The maximum rate of reaction is reached at 55°C (labelled B). Include the term 'optimum temperature' when you explain this section.

Section C shows the rate decreasing at temperatures above 55°C. Explain this by describing the effect of high temperatures on hydrogen bonds and how this affects the tertiary structure of the enzyme and the shape of the active site.

Focus on maths skills

Calculating the rate of an enzyme-controlled reaction using a tangent to a curve

You may be assessed on your ability to draw and use the slope (gradient) of a tangent to a curve as a measure of rate of change. The role of catalase in the decomposition of hydrogen peroxide was described in the background information section. The rate of this reaction can be found by measuring the volume of oxygen produced per unit time. The initial rate of reaction will be rapid because there are a lot of substrate molecules present and very little product. As the reaction proceeds, there will be fewer substrate molecules available and the rate of reaction will slow down. The initial rate of the reaction can be found using the following method:

1 Select the point at which you want to measure the gradient.
2 Draw a tangent to the curve at your chosen point. (A tangent is a straight line that just touches a curve, but does not cut across it.)
3 From your graph, find the change in the y-axis value and the change in the x-axis value.
4 Calculate the gradient using change in y divided by change in x.

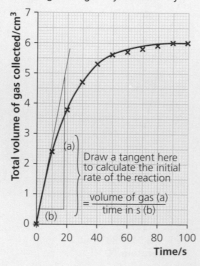

Figure 7 Calculating the initial rate of an enzyme-controlled reaction using a tangent to a curve

From Figure 7, the change in the y-axis value = $4.9 - 0.5 = 4.4\,cm^3$.

The change in the x-axis value = $18 - 2 = 16\,s$.

So the gradient = $4.4/16 = 0.275\,cm^3\,s^{-1}$.

This method is useful when investigating the rate of an enzyme-controlled reaction as it allows you to measure the change in rate as the reaction progresses.

Required practical 2

Preparing and observing a root tip squash

Background information

Eukaryotic cells that are able to divide show a cycle of cell growth and division called the **cell cycle**. **Interphase** takes up most of the cell cycle and is when DNA replication occurs. Interphase is followed by **mitosis**, when the cell divides to produce two identical daughter cells. Mitosis can be divided into four main stages

1 prophase   3 anaphase

2 metaphase   4 telophase

For each of these stages, you should be able to describe the behaviour of the chromosomes and explain the appearance of cells. You could copy and complete Table 2 and then use your textbook to prepare a summary before you complete the practical.

Table 2 A table that could be used to record the appearance of root tip cells

Stage of the cell cycle	Description of the cells' appearance	Explanation of the cells' appearance
Interphase		
Prophase		
Metaphase		
Anaphase		
Telophase		

Guidance through the practical

This practical requires you to complete three activities:

1 Prepare a slide — the tip of a plant root needs to be mounted on a slide, stained so that the chromosomes can be seen clearly, and squashed to produce a single layer of cells.

2 Use an optical microscope — you need to focus the microscope at a suitable magnification to allow individual cells to be seen and counted.

3 Calculate a mitotic index — both the number of cells undergoing mitosis and the total number of cells should be counted. The mitotic index is then calculated using the formula:

$$\text{mitotic index} = \frac{\text{number of cells undergoing mitosis}}{\text{total number of cells}}$$

Preparing the slide

To prepare the slide, you need access to plant roots — onion and garlic are commonly used in schools and colleges. The bulb can be stood on a beaker or conical flask full of water and the roots will be clearly visible after 2–5 days, as shown in Figure 8.

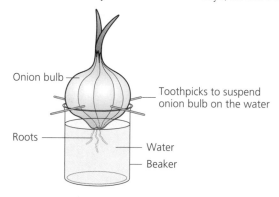

Figure 8 Onion suspended above water

It is important that you only use the very tip of the root, cutting approximately 1 mm off the end, because the **meristematic cells** found here are actively dividing. They should appear small and square when you look at them down the microscope, and the nucleus should be in the centre of the cell (Figure 9). The cells further away from the tip are no longer dividing; they are elongating. These cells will appear longer and the nucleus will not be central.

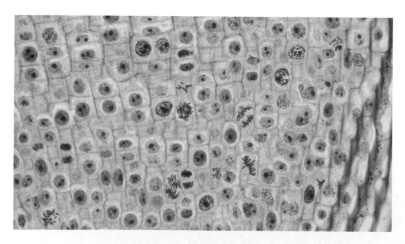

Figure 9 Meristematic cells from a root tip as seen using an optical microscope at medium power

Once the root tip has been removed, cell division must be halted immediately and the root tip prepared so that it can be observed using an optical microscope. The technique you use will depend on the availability of chemicals and a risk assessment carried out by the school or college, but the main steps are as follows:

1 **Fixing** — this kills the cells while causing minimum changes to them. Place the tissue in ethanoic alcohol and in concentrated hydrochloric acid. This stops cell division and also hydrolyses the **middle lamella** so that the cells can be separated easily.

The **middle lamella** is a layer of pectin that cements adjacent plant cells together.

2   **Staining** — place the root tip on a slide and add a stain that will bind to the chromosomes. This may be **toluidine blue**, which stains chromosomes blue, or **acetic orcein**, which stains chromosomes a purple-red colour.

3   **Preparing a root tip squash** — squash the stained root tip by placing a coverslip on top of the root tip and pressing down gently but firmly in the centre of the coverslip. This produces a single layer of cells, so you can see individual cells when you view the slide using a microscope, and allows the transmission of light through the specimen on the slide. Take care when doing this as the coverslip is made of glass and will break easily. Do not let the coverslip move from side to side, as this may damage or break the chromosomes.

Using an optical microscope

An optical microscope is also known as a light microscope because it relies on the transmission of light through the specimen on the stage. The microscopes used in schools and colleges vary, but they all have the same general structure shown in Figure 10.

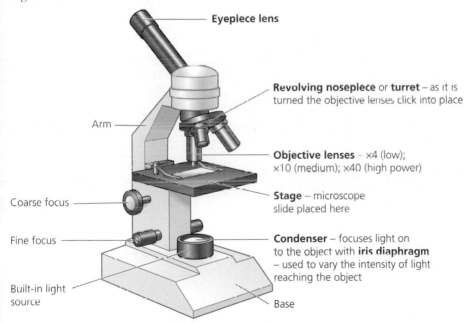

Figure 10 A typical optical microscope

When using a microscope, you should follow these steps:

1   Carry the microscope with one hand on the arm and the other under the base, and place the microscope on a stable surface in a well-lit area.

2   Turn on the light. You may need to plug the microscope in, or it may have a built-in battery.

3   Select the smallest **objective lens** by turning the turret until you feel the lens 'click' into place.

4 Place your slide on the **stage** and gently slide it under the clips, or into the slide holder if your microscope has a mechanical stage.

5 Looking at the stage from the side, use the **coarse focus** to either raise the stage or lower the lens (depending on the microscope) so that your slide is almost touching the objective lens.

6 Look through the **eyepiece lens**. The circular area that you can see is called the **field of view**.

7 Slowly adjust the coarse focus so that the slide is moving away from the objective lens.

8 Use the **fine focus** to bring the specimen into clear focus.

9 Turn the turret so that the medium objective lens clicks into place. Increasing the magnification decreases the field of view and fewer cells will be visible.

10 Look through the eyepiece lens and slowly adjust the **fine focus** until you can see the specimen clearly. Use a higher magnification if necessary. You are looking for small square cells arranged in rows, as seen in Figure 9.

Calculating a mitotic index

Once you have a clearly focused specimen, you should count the total number of cells that you can see. Record this number in your lab book with a clear description of what the number is — for example, total number of onion root tip cells seen at a magnification of ×100. You should then count the number of cells that are in any of the stages of mitosis (prophase, metaphase, anaphase or telophase) and record this number clearly in your lab book. Table 3 shows an example of a suitable table for recording your results.

Depending on the amount of time available, you could repeat these counts for several fields of view. Divide the number of cells undergoing mitosis by the total number of cells to calculate the mitotic index.

Table 3 An example of a table for recording your results and calculating mitotic index

Field of view	Total number of cells	Number of cells in stages of mitosis	Mitotic index
1			
2			
3			

Presentation of results from the practical

Your teacher may ask you to make cell drawings of your specimen in your lab book. Following the guidance on biological drawing on page 35, draw the cells to show the structures you can see at each stage of the cell cycle. Figure 11 shows both good and bad examples of biological drawings. Remember to label your drawing and include the magnification next to it. The maths skills box explains how to use an eyepiece graticule and calculate the magnification.

Practical tip

If you cannot see a clear image using the medium- or high-power objective lenses, select a lower-power (smaller) lens and refocus at the lower power.

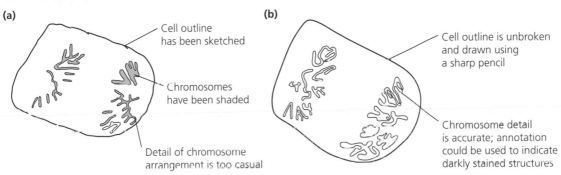

(a)
- Cell outline has been sketched
- Chromosomes have been shaded
- Detail of chromosome arrangement is too casual

(b)
- Cell outline is unbroken and drawn using a sharp pencil
- Chromosome detail is accurate; annotation could be used to indicate darkly stained structures

Figure 11 Biological drawings of a root tip cell undergoing mitosis. (a) this drawing makes many common mistakes; (b) an example of a clearly drawn cell

Analysis/evaluation of the results

In the exam, you might be given data about the duration of one cell cycle and the number of cells observed at each stage of mitosis. You could then be expected to calculate the period of time spent at each stage.

For example, in garlic root tissue, one cell cycle takes 8 hours. There are 50 cells and four of them are in metaphase. Calculate how many minutes these cells spend in metaphase.

The first step is to convert hours to minutes: 8 hours = 480 minutes.

The time spent in metaphase is:

$$\frac{4}{50} \times 480 = 38.4 \text{ minutes}$$

Focus on maths skills

Using an eyepiece graticule and calculating magnification

An eyepiece graticule is a glass or plastic disc with a scale on it. It is inserted into the eyepiece lens of a microscope and can be used to measure cells or structures. Before it can be used, it needs to be **calibrated** for each magnification (Figure 12).

Calibrating an eyepiece graticule:
1 Place a **stage micrometer** on the stage of a microscope. A stage micrometer is a glass slide with a scale marked at intervals, for example of 0.1 mm (100 μm) and 0.01 mm (10 μm).

→

2 Focus on the scale on the stage micrometer using the low-power objective lens.
3 Align the scales of the eyepiece graticule and the stage micrometer.
4 Count the number of divisions of the eyepiece graticule that are equivalent to 100 micrometres (100 μm) on the stage micrometer.
5 Calculate the length of one eyepiece division. For example, if 100 μm equals five eyepiece divisions, then each division equals 20 μm.
6 Repeat for the medium- and high-power objective lenses.

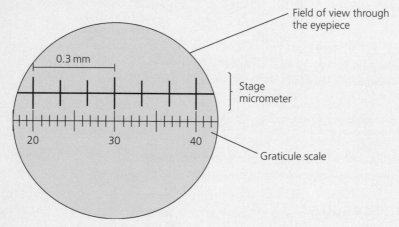

Figure 12 Calibrating an eyepiece graticule. The eyepiece is aligned with the stage micrometer. 10 divisions on the eyepiece graticule represent 0.3 mm, so each eyepiece graticule division = 0.03 mm or 30 μm

Using an eyepiece graticule to measure objects:
1 Place a prepared slide on the stage of a microscope.
2 Focus on the slide using the low-power objective lens.
3 Use the medium- or high-power objective lens to focus on an individual cell.
4 Measure the length of the cell by counting the number of eyepiece divisions.
5 Work out the **actual length** of the cell in micrometres using your calibrated values.

Calculating the magnification of a drawing:
1 Measure the length of the cell that you have drawn to give the **image length**.
2 Calculate the **magnification** using image length/actual length.
3 This magnification can be written next to your drawing.

Required practical 3

Using a calibration curve to estimate water potential

Background information

There are two elements to this practical — preparing a dilution series and then determining the water potential of plant tissue. Water potential is covered in the 'Transport across cell membranes' section of the specification, where you will have explained osmosis in terms of water potential. Water potential is measured in units of pressure (usually kilopascals, kPa) because it refers to the pressure created by water molecules. Pure water has a water potential of 0 kPa. The addition of any solute to this pure water will lower the water potential and therefore make it negative. Water always moves from higher water potential (less negative) to lower water potential (more negative).

The water potential of plant tissues can be determined by placing them in solutions with known solute concentrations. If the plant tissue *gains* mass, then water has *entered* the tissue by osmosis, and if the tissue *loses* mass, then water has *left* the tissue by osmosis. The concentration at which there is no loss or gain of mass is the one at which there has been no net movement of water by osmosis. This concentration has the same water potential as the plant tissue.

The specification does not name a particular plant tissue that should be used, but many schools and colleges use potatoes because they are readily available, are easy to cut to size, and give results within the time frame of a lesson. The solute is not specified either, but sodium chloride and sucrose are often used.

Guidance through the practical

Preparing a dilution series

A dilution series is prepared from a stock solution of known concentration, which is then diluted using distilled water. You may be told which concentrations to prepare, or you may have to decide on suitable concentrations for yourself, in which case you should choose even increments that will give you a minimum of 5 or 6 concentrations. For example, with a 1.0 mol dm^{-3} sucrose solution, you could prepare concentrations of 0.2, 0.4, 0.6 and 0.8 mol dm^{-3}. Distilled water on its own should also be used to give you a concentration of 0.0 mol dm^{-3}.

To prepare the dilution series you need to choose a piece of apparatus with a suitable scale to reduce the uncertainty of your readings. A burette with scale divisions of 0.1 cm^3 would be a more appropriate choice of apparatus than a measuring cylinder with divisions of 0.5 cm^3. The slightly curved surface of a liquid is called the **meniscus**. Take your reading from the bottom of the meniscus, as shown in Figure 13.

Once you have decided on the concentrations that you are going to use, you need to know the volumes of stock solution and distilled water to use. There are two steps to this process:

Step 1 Calculate the volume of stock solution. Use the equation:

$$\frac{\text{required concentration}}{\text{concentration of stock solution}} \times \text{final volume required}$$

Step 2 Calculate the volume of distilled water to add.

You then subtract the volume calculated in step 1 from the final volume required.

For example, if you wanted to prepare 30 cm^3 of 0.15 mol dm^{-3} sucrose solution from a stock solution of 0.5 mol dm^{-3} sucrose solution:

Step 1 $\frac{0.15}{0.5} \times 30 = 9$ cm^3 stock solution

Step 2 $30 - 9 = 21$ cm^3 distilled water

Make sure that each solution is poured into a boiling tube or other suitable container that has been clearly labelled with the concentration.

Exam tip

Pure water has a water potential of 0 and adding solutes makes it negative.

Exam tip

Data should always be recorded to a consistent number of decimal places, so the concentration of the distilled water is recorded as 0.0 mol dm^{-3} rather than just 0 mol dm^{-3}.

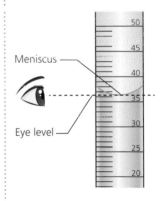

Figure 13 How to read a volume correctly from the bottom of a meniscus

This type of dilution is called **proportional dilution**. Another type of dilution that you might use is **serial dilution**. This can be carried out using different dilution factors. The method below describes how to dilute by a factor of 10:

1 Add $9\,cm^3$ of dilution liquid (often distilled water) to a series of numbered test tubes.

2 Transfer $1\,cm^3$ of stock solution to tube 1 using a graduated pipette.

3 Mix thoroughly by inverting the test tube. If the stock solution had a concentration of 10% then tube 1 contains a 1% solution as one-tenth of the total volume is stock solution (so 10%/10).

4 Transfer $1\,cm^3$ from tube 1 to tube 2 and mix thoroughly. Tube 2 now contains a 0.1% solution (1%/10).

5 Repeat the transfers until the required dilution series has been produced, as shown in Figure 14.

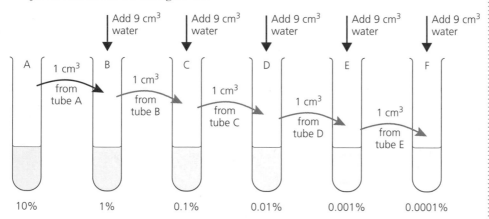

Figure 14 Making a serial dilution. Each transfer is reducing the concentration of glucose by a factor of 10

Once you have prepared the dilution series, you can move on to the next stage of the practical.

Determining the water potential of plant tissue

When preparing the plant tissue, it should be handled carefully to prevent damage to the cells. You should devise a method of preparing the plant tissue that ensures a uniform size, shape and surface area. Using a cork borer to produce potato cylinders is a straightforward technique that ensures a consistent diameter and the cylinders can then be cut to the same length. (If using potatoes you should make sure that the 'skin' (corky layer) is removed from the plant tissue.) Alternatively, you may be provided with a knife or a scalpel to cut the plant tissue to an appropriate size. Whether using a cork borer, knife or scalpel, the instrument should be used carefully to minimise the risk of injury (ATj).

The balance that you use to weigh the plant tissue should ideally measure to three decimal places, but two decimal places should give suitable results. The balance should be on a stable surface; most balances have a level indicator and adjustable legs so you can make sure that they are level.

Practical tip

Serial dilution can use dilution factors other than 10. To dilute by a factor of 2 you could transfer $5\,cm^3$ each time to produce a total volume of $10\,cm^3$.

Practical tip

Use a sharp blade and always cut away from yourself. Cut onto a suitable surface, for example a white tile or wooden board.

Record the mass of each piece of plant tissue as you weigh it, then place it directly into the required concentration and start your stopwatch.

After leaving the pieces of plant tissue in solution for the required period of time, gently blot them dry and weigh them again. You should take your lab book over to the balance with you and record the masses straight into your results table.

As with other investigations, you should identify the control variables and ensure that they remain constant. The significant variables that need controlling include:

- temperature
- source of plant tissue — for example, use the same potato
- size and shape of plant tissue
- batch of stock solution
- length of time the plant tissue is immersed in the solution

For each of these, you should be able to explain how and why you are controlling them.

Presentation of results from the practical

You should draw your results table in your lab book before starting the practical, ensuring that you have sufficient columns and space for all of the raw data you plan to collect. You can always add columns if necessary, or draw a second table for processed data. A table for **raw data** would need to include columns for:

- the concentration of solute — the units for this may be moles per cubic decimetre ($mol\,dm^{-3}$), which can also be written as molarity (M)
- the initial mass of the plant tissue/g
- the final mass of the plant tissue/g

You will then use your raw data to produce your processed data. Processed data are calculations using your raw data and will include:

- the change in mass/g (final mass minus initial mass)
- the percentage change in mass (see page 29)

You should also convert the concentrations of the sucrose solution to water potential using Table 4.

Table 4 Conversions of sucrose concentration to water potential

Concentration of sucrose solution/$mol\,dm^{-3}$	Water potential/kPa
0.00	0
0.20	−540
0.40	−1120
0.60	−1800
0.80	−2580
1.00	−3500

Your processed data should then be presented as a line graph because the variables are continuous. Figure 15 shows an example of the type of graph you could plot.

The graph should have:

- water potential on the x-axis and the percentage change in mass on the y-axis
- a y-axis scale that allows you to plot both positive and negative values

- a suitable scale, with the same scale divisions on both the positive and negative sections of the *y*-axis
- points joined with a suitable line

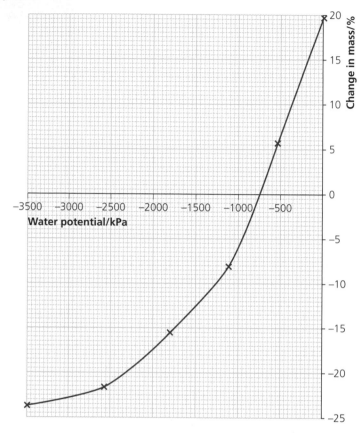

Figure 15 Graph showing the effect of water potential on the percentage change in mass of plant tissue. The *x*-axis intercept is equal to the water potential of the plant tissue (–750 kPa)

Analysis/evaluation of the results

Once you have plotted your graph, you will notice that there are three main sections:

- Above the *x*-axis where there is an increase in mass.
- The *x*-axis intercept, where there is no overall change in mass.
- Below the *x*-axis, where there is a decrease in mass.

For each of these sections, you should be able to explain what is happening to the plant tissue in terms of the loss or gain of water by osmosis, and whether the water potential of the solution is less negative or more negative than the plant tissue.

You can use your graph to determine the water potential of the plant tissue using the *x*-intercept. Find the point where the line crosses the *x*-axis and record the water potential. At this point, there is no change in mass and therefore no net movement of water into or out of the plant tissue.

Exam tip

Remember that when the solute concentration is *low*, the water potential is *high*, so water enters the cells by osmosis and the plant tissue *gains* mass.

Referencing

Remember that all sources should be credited when writing a report in your lab book, to provide evidence for CPAC 5.

If a website is used, you must include the full web address, plus the date and time of access. For example:

> http://www.nuffieldfoundation.org/practical-biology/investigating-effect-concentration-blackcurrant-squash-osmosis-chipped-potatoes (21/12/2016; 17:15)

If a book is used, you must include the author's name, year of publication, title, edition, city where published and name of publisher. For example:

> Lowrie, P., Smith, M., Bailey, M., Indge, B. and Rowland, M. (2015) *AQA A-level Biology 1,* 1st edition, London: Hodder Education.

Focus on maths skills

Percentage change

Calculating percentage change is important because it allows us to compare the change in mass of plant tissue samples that had a different initial mass. It is calculated by:

$$\frac{\text{change in mass}}{\text{initial mass}} \times 100$$

You should include a plus (+) sign to indicate an increase in mass and a minus (–) sign to show a decrease in mass.

You may be assessed on your ability to calculate percentage change, percentage increase and percentage decrease. The principle is always the same:

$$\frac{\text{difference}}{\text{original}} \times 100$$

Exam tip

Remember that percentage increase can be greater than 100% because it is possible to have more than a 100% increase. For example, a crop yield that has increased from 80 kg to 190 kg shows a 137.5% increase.

Required practical 4

Factors affecting membrane permeability

Background information

In this practical, you will investigate the effect of a named variable on the permeability of cell-surface membranes. The cell-surface membrane refers to the plasma membrane that forms the boundary between the cytoplasm and the external environment. Membranes are also found around and in cell organelles, and all have the same basic structure.

The fluid-mosaic structure of cell membranes is covered in the 'Transport across cell membranes' section of the specification, and an important synoptic link to make is between the presence of proteins in membranes and the properties of proteins. Any factor that affects the tertiary structure of a protein, such as temperature or pH, will affect the permeability of the membrane. Similarly, factors that change the fluidity of the phospholipid bilayer of a membrane will affect its permeability. High temperatures and extremes of pH can denature the proteins in cell membranes, leading to an increase in permeability. Some lipid-soluble solvents, such as alcohol, can damage cell membranes, again leading to an increase in permeability.

Any investigation that you carry out will require you to take measurements or make observations about membrane permeability. For that reason, the techniques used generally involve using plant tissues that contain coloured pigments, such as beetroot or red cabbage. Both beetroot and red cabbage contain water-soluble pigments that are in solution in the cell vacuole. When the membranes of these plant tissues are more permeable, the coloured pigment can leak out more readily through the fully permeable cell walls and diffuse into the surrounding solution. Both the cell-surface membrane *and* the **tonoplast** — the membrane around the vacuole (Figure 16) — must be disrupted for the pigment to leave the vacuole. The intensity of the colour of the surrounding solution is directly related to membrane permeability and could be measured **semi-quantitatively** by comparing the solutions to colour standards, or **quantitatively** using a colorimeter.

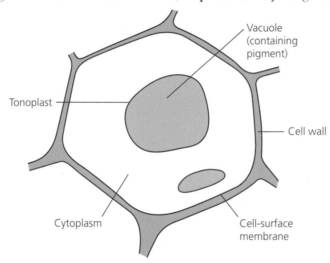

Figure 16 A plant cell, showing the structures through which pigments must diffuse

Guidance through the practical

The independent variable investigated is often temperature, because as there is a clear relationship between temperature and membrane permeability there is a good link to the proteins section of the specification. You may also investigate the effect of different types of alcohol or of different concentrations of an alcohol, for example ethanol, on membrane permeability.

The availability of equipment will affect your choice of dependent variable, because a colorimeter is required to obtain quantitative data.

If you investigate the effect of temperature, important variables to control would include:
- source of plant tissue
- time in solution
- size and shape of plant tissue

Any plant tissue that is used must be fresh; preservation techniques, such as pickling or canning, would have damaged the cell membranes. Your method of cutting the plant tissue will depend on the type of plant you use, but with beetroot a cork borer

Practical tip

Quantitative results are numerical values, whereas qualitative results are non-numerical or descriptive.

Semi-quantitative results allow you to estimate the amount or concentration of a substance.

can be used to cut out cylinders and then these cylinders can be cut into equal-sized discs. When you cut the plant tissue, particularly if you use beetroot, some pigment will be released so you will need to rinse the discs thoroughly using distilled water. Care should be taken when using cork borers, scalpels and knives.

Method 1 Semi-quantitative technique

Part 1

For this method, the first step is preparing a series of colour standards. You will then use these to compare your test solutions against when investigating membrane permeability. You need to start with distilled water and a solution of pure plant extract, for example beetroot extract, and then use these to prepare a series of different concentrations at equal intervals. The distilled water will be 0% and the pure plant extract will be 100%, so your dilutions may be 20%, 40%, 60% and 80%.

Table 5 shows the volumes of water and plant extract that you could use to make 10 cm³ of each concentration of extract.

Table 5 How to dilute a plant extract using proportional dilution

Concentration of extract/%	0	20	40	60	80	100
Volume of distilled water/cm³	10	8	6	4	2	0
Volume of plant extract/cm³	0	2	4	6	8	10

Once you have a series of colour standards, you can investigate your chosen variable. This example will now focus on temperature.

Part 2

1 Set up water baths at the required temperatures, for example, 20°C, 30°C, 40°C, 50°C and 60°C. Ideally these will be thermostatically controlled water baths, but if these are not available then a beaker of water can be used. It is important to monitor the temperature throughout the investigation using a thermometer checked at regular intervals. Hot water can be added to the water bath if the temperature starts to fall.

2 Label clean test tubes with each temperature.

3 Measure the same volume of distilled water into each tube.

4 Place the same number of beetroot discs, or other suitable plant tissue, into each tube.

5 Leave the tubes in the water bath for the same period of time.

6 Gently shake the tubes at regular intervals to maintain the diffusion gradient for the pigment.

7 At the end of the time in the water bath, remove the beetroot discs so that no more pigment diffuses out.

8 Compare the colour of the solution from each temperature with your range of colour standards and record the percentage concentration in a suitable results table.

There are two main issues with this method:
- Matching to the colour standards is **subjective**.
- The colour you obtain may not match any of your colour standards.

Subjective results depend on the interpretation or judgement of the person making the observation.

Method 2 Quantitative technique using a colorimeter

If a colorimeter is available for use, the issues associated with using colour standards do not apply. You can just carry out part 2 of the method described above so that you have test tubes containing coloured solution from each temperature. A colorimeter can be used to measure the absorbance of each solution, and this value can be recorded in your results table.

Using a colorimeter

A colorimeter is an instrument that can be used to determine the concentration of solute in a solution. It does this by shining light through a solution and measuring either the **transmission** (the percentage of light that passes through the solution) or the **absorbance** (the light absorbed by the solution). The concentration of the solute is proportional to the absorbance of the solution, so a more concentrated solution will have a higher absorbance.

Colorimeters have coloured filters and can be adjusted to select a suitable wavelength of light. You should choose the colour that is complementary to the colour of your solution, as this will give the maximum absorbance. Complementary colours are opposite each other on the colour wheel, as shown in Figure 17, so you should select a green filter if your solution is red.

Some colorimeters require you to use a **cuvette** (Figure 18), whereas others are designed to accommodate both test tubes and cuvettes.

> **Practical tip**
>
> A colorimeter can measure absorbance or percentage transmission.

> A **cuvette** is a clear, straight-sided container made of glass or plastic.

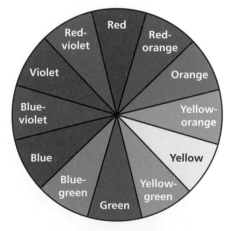

Figure 17 A colour wheel

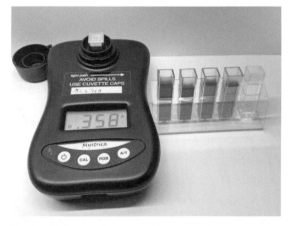

Figure 18 A colorimeter with a sample in a cuvette

Before taking readings with the colorimeter, you need to set it to give an absorbance reading of zero (or a percentage light transmission of 100%). The solution that you use to do this is called a 'blank' and will be distilled water in this practical. When using a colorimeter, you are looking for a change in colour, so the blank is not always distilled water.

Presentation of results from the practical

results table should be drawn in your lab book before you start the practical. The ...ent variable, for example temperature, should be in the first column and ...nt variable in the following column. The dependent variable may be your

> **Practical tip**
>
> Absorbance does *not* have units because it is a ratio (of the amount of light that passes through a solution compared with the amount of light that passes into the solution).

estimation of pigment concentration if you used colour standards, or your reading of absorbance or light transmission from the colorimeter. Remember to include units in the column headings only.

You should plot a graph of your results with the factor that you changed on the *x*-axis and either the estimated pigment concentration or the absorbance on the *y*-axis. The graph will be a line graph if you have investigated temperature, pH or alcohol concentration, as these are all continuous variables. Remember to join the points with straight lines if you cannot be certain of the intermediate values.

Analysis/evaluation of the results

In your analysis of your results, you should *describe* any trend or pattern that you can see, using values to illustrate this.

You should then *explain* the pattern or trend using your scientific knowledge. If the absorbance of a solution is *high* (or the percentage light transmission is *low*), a lot of the light entering the solution has been absorbed, meaning that the solution is darker, so in this investigation more pigment has been released. For more pigment to pass into the solution, both the cell-surface membrane *and* the tonoplast must be more permeable. You should explain how the factor you are investigating has increased this permeability.

Focus on maths skills

Linear relationship

You should understand that the equation below represents a linear relationship:

$$y = mx + c$$

where y = the dependent variable, m = the gradient of the line, x = the independent variable and c = the intercept on the y-axis.

If $c = 0$, the line goes through the origin, so that y is directly proportional to x. This relationship is shown in Figure 19 for the effect of enzyme concentration on the rate of reaction.

You can calculate the rate of change by finding the gradient of the line $\left(\dfrac{dy}{dx}\right)$.

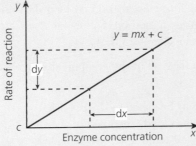

Figure 19 Graph showing the effect of enzyme concentration on rate of reaction (where substrate is not a limiting factor)

Required practical 5

Dissection of an organ or system from an animal or plant

Background information

The organ or system that you will dissect is not prescribed by AQA, leaving the choice to individual schools and colleges. You could dissect an entire gas exchange system or mass transport system, or you may dissect one organ from within the system, such as a heart.

The specification covers mass transport in both mammals and plants, and gas exchange in insects, fish and dicotyledonous plants. You must be familiar with the features of the mass transport or gas exchange system before you carry out your dissection.

Guidance through the practical

This practical gives you the opportunity to develop two important skills:
- Safely using instruments for dissection (ATj).
- Producing a scientific drawing with annotations (ATe).

If you are using an animal organ or system, the most important factor is the freshness of the material. Students are often put off dissection because of the smell, but fresh organs will not smell very strong. Remember that any animal material that you use should be kept in a sealed container and stored in the fridge. If you are storing it for more than a day, it can be frozen and then defrosted overnight in the fridge prior to use.

The most common dissections in schools and colleges are hearts or fish heads, as these are cheap and readily available. Hearts can be bought pre-packed from supermarkets but these tend to be 'trimmed', which means that the blood vessels and atria have been removed, whereas a butcher may be able to supply an 'untrimmed' heart. Fish heads are often given away free of charge from fish counters or market stalls.

Other possible dissections are the lungs of a mammal, the gas exchange system of a large insect such as a locust, or a plant mass transport system. The freedom to choose to dissect a plant organ or system makes this practical accessible to all students, including those who prefer not to use animals.

A dissection must be planned and carried out methodically, as its aim is to study the anatomy of the organ or system rather than just hacking it to pieces. You may be given a method to follow or you may be asked to research your own method. Following written instructions is an important skill and provides evidence for CPAC 1.

Preparation is essential and you should ensure that you have all of the equipment you need prior to beginning your dissection.

Both safety and hygiene must be considered, and you may be asked to write a risk assessment before beginning the practical. This should be clearly recorded in your lab book and, together with your teacher's observations of your technique, will provide evidence for CPAC 3. A risk assessment is often recorded in a table, with hazards, risks and control measures identified. Table 6 shows a student's risk assessment for a heart dissection.

Table 6 Hazards, risks and control measures associated with carrying out a heart dissection

Hazards	Risks	Control measures
Scalpel	You might cut yourself	Use a sharp, rust-free blade Cut away from yourself
Handling the heart	You might contaminate the work bench There may be bacteria on or in the heart	Wipe down the bench with disinfectant at the end Wash hands thoroughly at the end of the practical

The following safety precautions should be taken when dissecting an animal organ or system:

- You should wear a disposable plastic apron or lab coat.
- You could wear disposable gloves if they are available.
- Any cuts or open wounds should be covered with waterproof dressings.
- Scalpels and other instruments should be handled with care.
- Materials such as organs and plastic gloves should be disposed of safely.
- All instruments and surfaces should be disinfected at the end of the session.
- Hands should be washed thoroughly with soap and warm water.

Presentation of results from the practical

This practical provides you with an excellent opportunity to present your results in a different format instead of the tables and graphs that are commonly required. Photographs of your dissection can be stuck directly into your lab book, then labelled and annotated. Alternatively, you could paste the photo into a word document then label and annotate before printing it off and sticking it in your lab book. You can also develop your skills of observation by accurately drawing the structures and tissues that you can see (Figure 20). There are a number of rules to follow with biological drawing:

- Use a sharp pencil.
- Do not shade or use colour.
- Use unbroken lines; do not sketch.
- Label the structures and include brief annotations.
- Include a scale or the magnification.

There are three main types of biological drawing you may complete during the A-level course:

- **Cell drawings** show the components of individual cells as observed using an optical microscope.
- **Tissue maps** show the location of tissues in an organ or organism.
- **Body plans** show the main body parts of a dissected organism.

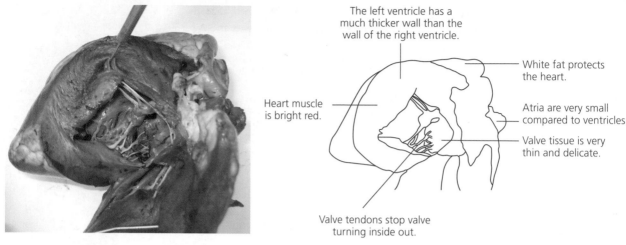

Figure 20 A photo showing a section of a heart, together with an annotated biological drawing

The left ventricle has a much thicker wall than the wall of the right ventricle.

White fat protects the heart.

Heart muscle is bright red.

Atria are very small compared to ventricles.

Valve tissue is very thin and delicate.

Valve tendons stop valve turning inside out.

Analysis/evaluation of the results

Once you have finished your dissection, you should write a description of the structures that you observed and relate these to your biological knowledge. For example, with a heart dissection you could describe and explain the relative thickness of the chamber walls, the shape of the valves or the strength of the tendons. With a mammalian lung dissection, you could describe and explain the spongy texture of the lungs, the ridged texture of the trachea and bronchi, and differences between the pulmonary artery and pulmonary vein.

Required practical 6

Using aseptic techniques to investigate the effect of antimicrobials

Background information

This practical requires you to investigate the effect of an antimicrobial substance on microbial growth. An antimicrobial substance is any chemical that either kills a microbe or inhibits its growth. Antimicrobials include:

- disinfectants, for example bleach, which can be used on non-living surfaces to kill microbes
- antiseptics, for example ethanol, which are applied to living tissue or skin to reduce the risk of infection
- antibiotics, for example penicillin, which are used to destroy microbes in the body

Microbe is a term that can apply to any microscopic organism, including bacteria, fungi, parasites and viruses.

Antibiotics are a group of chemicals that are used to treat or prevent bacterial infections. There are many different antibiotics with different mechanisms of action — for example, penicillin inhibits cell wall synthesis in growing bacteria.

Testing bacteria for their sensitivity to antibiotics allows doctors to prescribe the correct antibiotic at an effective dosage. This is important in minimising the rise of antibiotic-resistant bacteria.

Exam tip

Antibiotic resistance arises from a random mutation that leads to a new allele of a gene. As this new allele is advantageous, it is inherited by the next generation and the frequency of the allele increases in the population.

Guidance through the practical

This practical gives you the opportunity to use aseptic technique to culture bacteria, and to use a range of apparatus and materials including **agar plates** and **nutrient broth** (AT i).

Agar plates are sterile plastic Petri dishes that have had molten nutrient agar poured into them and are used for culturing bacteria. Nutrient agar is supplied in powder form and is mixed with distilled water and heated before use. The molten agar solidifies as it cools to form a jelly-like substance and microorganisms can grow on the surface. Nutrient agar contains the carbohydrates, organic nitrogen (in the form of peptone — partly digested protein), vitamins and sodium chloride required for the growth of a wide range of microbes, as well as agar powder.

Nutrient broth is a liquid growth medium that contains the nutrients needed for bacterial growth, but does not contain agar. Bacteria can be cultured in small glass bottles containing nutrient broth.

Aseptic technique refers to the methods used to prevent contamination by microbes. There are many factors to consider and so your teacher may demonstrate these techniques to you.

- Ensure that the bench or work surface has been thoroughly cleaned and disinfected.
- Light a Bunsen burner and work near the lit Bunsen throughout the practical; the warm air current will create an updraft that reduces the risk of air-borne microbes contaminating your agar.
- Use sterile equipment.

Disposable loops, spreaders and pipettes may be provided in sterile, sealed packages. Alternatively the technician may have sterilised equipment by **autoclaving** (heating at high temperature and pressure).

A Bunsen burner can also be used to sterilise equipment: metal loops can be held in the hottest part of the flame until they glow red hot; glass spreaders can be dipped in ethanol and then passed through the flame to allow the alcohol to burn off.

Remember to let the equipment cool down before it comes into contact with the bacteria, but do not put sterilised equipment down on the bench before use or it may become contaminated.

Your school or college will have chosen the bacteria that you use — they must be easy to culture at low temperatures and should present no risk to health. Common choices include *Bacillus megaterium*, *Escherichia coli* (*E. coli*) and *Micrococcus luteus*.

Before removing a sample of bacteria in broth from the bottle, you may be told to 'flame the neck of the bottle'. This means that you should remove the lid, pass the neck of the bottle through the flame, remove your sample using a sterile pipette, pass the neck of the bottle through the flame again and replace the lid. The bottle does not need to be held in the flame for a long period of time. This process of flaming warms the air around the neck of the bottle so that movement of air is out of the bottle, which prevents contamination of the broth by microbes in the air.

When adding the bacteria to your agar plate, the lid of the Petri dish should be lifted as little as possible. You should then use a sterile spreader to give an even covering

Practical tip

Nutrient agar is just one of the many types of agar used for culturing microorganisms. Agar, which is extracted from algae, can be added to different liquid growth media to produce selective, indicator or enriched agar plates.

Practical tip

Care must be taken if using ethanol near an open flame because it is flammable.

Practical tip

Note that the names of the bacteria are binomials and represent the genus and the species. The genus begins with an upper-case letter, and the species with a lower-case letter. In print, the names are *italicised*, but when handwritten in your lab books, they should be underlined.

over the surface of the agar. After incubation, this will give a continuous layer of bacteria called a **bacterial lawn**.

You should use sterile forceps to place your antimicrobial agent on the agar plate, again lifting the lid as little as possible. You might use a multidisc, which is a ring with different antimicrobial agents, or individual paper discs impregnated with antibiotics. Alternatively, you could make your own paper discs using sterile filter paper soaked in different concentrations of bleach.

The lid of the Petri dish should be held in place with two pieces of tape. It should not be sealed all the way around as this may favour the growth of potentially harmful anaerobic bacteria. The plate should be incubated upside down in an incubator at no more than 25°C. Inverting the plates during incubation means that condensation drips onto the lid rather than on the surface of the agar. Keeping the temperature of the incubator well below human body temperature reduces the risk of growing microbes that are pathogenic to humans.

Incubate the plate for 48 hours then, *without removing the lid* of the Petri dish, examine the growth of the bacteria. You should see clear zones around one or more of the discs where bacterial growth has been inhibited. These clear zones are called **zones of inhibition** (Figure 21). The larger the zone, the more effective the antimicrobial. The radius of each clear zone should be measured and recorded in a suitable results table.

Figure 22 shows the effect of eight antibiotics in a multidisc (mast ring) on the growth of bacteria. Different antibiotics are contained in the arms of the mast ring, so that sensitivity to many antibiotics can be tested simultaneously.

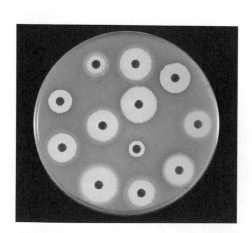

Figure 21 Photograph showing bacteria growing on an agar plate, with zones of inhibition of varying diameters around the antibiotic discs

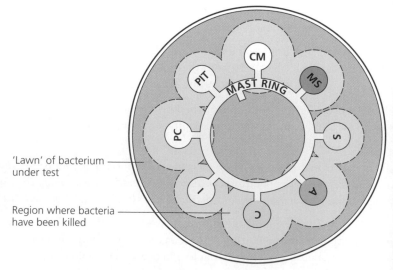

'Lawn' of bacterium under test

Region where bacteria have been killed

Figure 22 Diagram showing how a multidisc (or mast ring) with different antibiotics affects bacterial growth. The 'clear' areas are zones of inhibition of varying sizes

It is important to disinfect the benches and dispose of agar plates safely at the end of the practical session. Hands should be washed thoroughly with soap and warm water after handling microbes.

Presentation of results from the practical

You could take a photograph of your plate and stick it in your lab book, or you could draw a diagram of your plate, ensuring that it is an accurate record of your observations.

A table should be drawn with the name of each antibiotic in the first column and the radius of each zone of inhibition in the next column. As each zone of inhibition may not be uniform, you should take several readings and calculate a mean. Measurements should be recorded in millimetres, and you should take each measurement from the centre of the disc. It may be easier to determine the mean diameter and then use this to calculate the radius.

You can add a further column for processed data and use the mean radius to calculate the area of the zone of inhibition for each antibiotic. Remember to include suitable units for area in the column heading (probably mm^2).

You could display your results graphically.

- If you have used different antibiotics, you should draw a bar chart with the type of antibiotic on the x-axis and the area of the zone of inhibition on the y-axis. The bars should not be touching each other, as different antibiotics are an example of categorical data.
- You may have used **serial dilution** to prepare different concentrations of one antimicrobial. In this case you should draw a line graph, as concentration is an example of continuous data, with concentration on the x-axis and area of the zone of inhibition on the y-axis.

Analysis/evaluation of the results

You should explain your observations and measurements using your scientific knowledge, and may have to use different sources to research the antibiotics and microbe that you used. If you have used different antibiotics there may be zones of inhibition around some of the discs but not around others, so your analysis should suggest why this occurred. Antibiotics have different mechanisms of action, so may not be effective against the microbe that you used. Alternatively, the lack of a zone of inhibition could indicate antibiotic resistance. You could also consider the physical properties of the antibiotic in terms of the size of the molecules and how this affects their rate of diffusion through the agar.

Required practical 7

Using chromatography to investigate leaf pigments

Background information

The leaves of plants contain photosynthetic pigments that absorb light energy and transfer it to chemical energy in the molecules formed during photosynthesis. These photosynthetic pigments are found in the chloroplasts of eukaryotic cells, embedded in their thylakoid membranes.

> **Exam tip**
>
> You may be tested on your ability to calculate the area of a circle using πr^2 and the circumference of a circle using $2\pi r$, where r = radius and π = 3.14.

> Categorical data relate to category, such as qualitative data, as opposed to numerical data — for example, names of bacteria or types of antibiotic.

There are many different pigments found in plants, but they all reflect certain wavelengths of light, making them appear coloured — for example, chlorophyll reflects green light. Leaves can contain different pigments, including:

- chlorophyll a — a blue-green pigment that absorbs violet/blue and red light
- chlorophyll b — a yellow-green pigment that absorbs blue and orange/red light
- carotenoids — yellow or yellow-orange pigments that absorb violet, blue and green light
- anthocyanins — red, purple or blue pigments that are not involved in photosynthesis

The carotenoids, including carotene and xanthophyll, are known as accessory pigments and absorb wavelengths of light that chlorophyll cannot.

Paper chromatography can be used to separate a mixture of molecules, for example a mixture of plant pigments. A solvent called the **extraction solvent** is used to extract pigments from a leaf and then a **running solvent** is used to draw the pigments up a piece of filter paper. As the running solvent travels up the filter paper, the pigment molecules are carried up the paper in solution. The rate at which they are carried depends on the solubility of the pigments, the size of the pigment molecules, and the affinity of the pigments for the filter paper (their tendency to adhere or 'stick' to the filter paper).

The filter paper should be left in the solvent for long enough for the pigments to be separated and until the solvent almost reaches the top of the filter paper.

Guidance through the practical

This practical gives you the opportunity to use paper chromatography (AT g), and the full title of the practical specifies that you should investigate leaves from different plants. One of the suggestions is that you compare leaves from shade-tolerant and shade-intolerant plants. Shade-tolerant plants are those that are able to photosynthesise at low light intensities, such as the European beech, whereas shade-intolerant plants like the silver birch need very high light intensities.

In a woodland, most of the red and blue light is absorbed by the shade-intolerant canopy plants, but far-red light (about 730 nm) and green light can penetrate the canopy. Shade-tolerant plants need photosynthetic pigments that allow them to maximise the absorption of light of these wavelengths.

You could also compare leaves that are different colours. Some leaves appear different colours throughout the year, like those of *Begonia*, which are red, and *Oxalis*, which are purple, or leaves could be collected in the autumn and the pigments present compared.

Figure 23 shows one way that paper chromatography could be set-up.

The first step in this practical is to set up boiling tubes containing a small volume of the running solvent that you are using and seal them with bungs. This will allow the air inside the tubes to become saturated with the solvent.

You should then collect a strip of filter or chromatography paper and draw a *pencil* line called the **origin** about 2 cm from the bottom. You must use pencil as ink may be soluble in the solvent that you use. The chromatography paper should be narrow

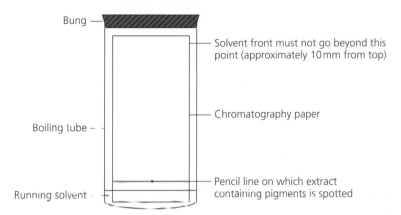

Figure 23 Diagram showing how to set up paper chromatography

enough not to touch the sides of the boiling tube as this may cause the solvent front to travel at an angle.

The next step is extracting the pigments. This can be achieved by grinding the leaf with a small amount of extraction solvent, such as propanone, using a mortar and pestle. A capillary tube can be used to spot the pigment extract onto the origin line. This is most effective if the paper is allowed to dry between applications to build up a concentrated spot of pigment. Alternatively, the pigments can be transferred directly from the leaf onto the filter paper by placing a leaf disc on the origin line and firmly crushing it with a glass rod.

You should then place the chromatography paper in the boiling tube, making sure that the end of the paper touches the solvent, but that the solvent is below the origin line. If the solvent is above the origin, the pigment will just be 'washed off' the paper. The bung should be replaced and the boiling tube left in a boiling tube rack.

The solvent will gradually rise up the paper and should be left until it *almost* reaches the top. You then need to remove the paper from the solvent and use a pencil to immediately draw a line to mark how far the solvent has travelled. This line is called the **solvent front**. You should do this immediately because you will not be able to see the solvent front once the chromatography paper has dried.

Your chromatography paper with the pigment spots is called a **chromatogram**.

Presentation of results from the practical

Your chromatogram should be left to dry and then can be stuck in your lab book. Use a pencil to draw around each pigment spot, as they may fade with time. You should also number each pigment to make it easier to refer to in your analysis.

Draw a results table for each chromatogram, with the number allocated to the pigment spots in the first column, their colour in the second column and the distance from the origin to the centre of the pigment spot recorded in mm in the final column.

Process your data by calculating the Rf value for each pigment using this formula:

$$\text{Rf value} = \frac{\text{distance from the origin to the centre of the pigment spot}}{\text{distance from the origin to the solvent front}}$$

These Rf values can be recorded in an additional column in your results table.

Analysis/evaluation of the results

The Rf value is a ratio of the distance travelled by the pigment and the distance travelled by the solvent. Calculating this value allows you to compare results with others because the ratio for a pigment will be the same even if the actual distances are different. You could identify the pigments on your chromatogram by researching the expected Rf values for each pigment. If there are differences between your values and the expected values, you could suggest possible reasons for these differences.

You should have chromatograms for at least two different leaves and should carry out some research to explain why the leaves contain different pigments.

Required practical 8

Factors affecting dehydrogenase activity in chloroplasts

Background information

In this practical, you are measuring the rate of the **Hill reaction** in isolated chloroplasts. Robert Hill carried out this investigation in 1938 and concluded that water had been split into hydrogen and oxygen.

Before beginning this practical, you need a clear understanding of both the light-dependent and the light-independent reactions of photosynthesis. The light-dependent reaction involves chlorophyll molecules, which are located in the thylakoid membranes, absorbing light energy and losing electrons in a process called **photoionisation**. The electrons are taken up by an electron carrier and then passed along a series of carriers in an electron transfer chain. The electrons lose energy as they pass along the chain to each carrier, and some of this energy is used to synthesise ATP.

Light energy is also used to split a molecule of water in a process known as **photolysis**. This produces protons, electrons and oxygen. Protons and electrons are taken up by an electron carrier called NADP, therefore **reducing** the NADP. The oxygen is used in respiration or diffuses out of the leaf through the stomata.

In summary, the light-dependent reaction produces ATP and reduced NADP, together with oxygen (from the photolysis of water). The ATP and reduced NADP are used in the light-independent reaction of photosynthesis.

DCPIP (2,6-dichlorophenol-indophenol) is a blue dye that acts as an electron acceptor. DCPIP is blue when **oxidised**, and accepting electrons **reduces** the DCPIP, causing it to decolourise. If DCPIP is added to a suspension of isolated chloroplasts, the DCPIP accepts electrons from the electron transfer chain. This reduces the DCPIP, so it changes colour from blue to colourless.

Guidance through the practical

This practical requires you to investigate the effect of a factor on the activity of chloroplasts. The thylakoids in chloroplasts contain pigments that absorb light, and proteins that are involved in the transfer of electrons. These proteins form the electron transfer chain, which consists of a number of electron carriers embedded in

Exam tip

NADP is an electron carrier in photosynthesis. Do not mix this up with NAD, which is involved in respiration.

Exam tip

Remember **OIL RIG**:

Oxidation **is** **l**oss of electrons or hydrogen (or gain of oxygen).

Reduction **is** **g**ain of electrons or hydrogen (or loss of oxygen).

the thylakoid membranes. You can investigate any factors that affect the pigments, or any that influence the components of the electron transfer chain.

During this practical, you will choose one factor as your independent variable and ensure that the other factors are controlled.

You can investigate the effect of light intensity and light wavelength on the rate of photosynthesis using the Hill reaction. Some weedkillers act on chloroplasts by inhibiting components of the electron transfer chain and therefore preventing the production of ATP and reduced NADP. Ammonium hydroxide has a similar effect and could be used to investigate the effect of these weedkillers on the rate of photosynthesis.

The effect of light intensity could be investigated by placing your reaction tube at different distances from a lamp. The effect of different wavelengths of light could be investigated by placing filters of different colours (e.g. blue, green and red) between the lamp and the reaction tubes, or wrapping the tubes in the different filters.

Weedkillers could be investigated using ammonium hydroxide to mimic their effects. In theory, the effect of different weedkillers could be investigated, or the effect of a weedkiller at different concentrations. However, a full risk assessment would be necessary to determine whether weedkillers could be safely used in the laboratory.

A source of chloroplasts needs to be selected and the plant material prepared. Spinach leaves are a good choice as they are readily available. These should be prepared by removing the stalks and midribs, which are too tough to blend and may not contain many chloroplasts.

You will need to isolate chloroplasts from the plant material by blending the leaves with ice-cold buffer solution and then filtering the solution. The blending breaks open the cell walls, releasing the cell contents, and filtering removes the cell walls and other debris to leave a suspension of chloroplasts. The buffer needs to be ice-cold to slow down enzyme activity and reduce damage to the chloroplasts. The buffer should also have the same water potential as the cytoplasm so that there is no net movement of water into or out of the chloroplasts by osmosis.

A small volume of your chloroplast suspension can be measured into a test tube and DCPIP added. You can then measure the time taken for the DCPIP to decolourise in different conditions, depending on the independent variable that you chose.

Variables that you will need to control include:
- source of chloroplasts — use the same suspension throughout your investigation
- volume of chloroplast suspension
- volume of DCPIP
- concentration of DCPIP

You will need to ensure consistency when determining the end-point of the reaction and should prepare a colour standard to determine when the DCPIP has completely decolourised. This will be a tube containing just chloroplast suspension and water (use the same volume of water as DCPIP).

Practical tip

If the colour change takes more than 2–3 minutes the DCPIP can be diluted.

Skills Guidance

Presentation of results from the practical

All results should be recorded in a suitable table with full column headings and units in the column headings only. When recording time, you should never have mixed units, for example, minutes and seconds. Convert the minutes into seconds and consider whether it is appropriate to include hundredths of a second.

As you have measured the time taken for DCPIP to decolourise, you can calculate the rate using 1/time, with a unit of s^{-1}.

Plot a graph of your results with the factor that you investigated, such as light intensity or concentration of ammonium hydroxide, on the x-axis and rate of photosynthesis on the y-axis. These are all continuous data, so your graph will be a line graph. Draw a bar chart if you have used different inhibitors or different filters (if you do not know the wavelength of light that passes through each one).

Analysis/evaluation of the results

You should describe any patterns or trends shown by your results, and then explain these using your scientific knowledge. Remember that the faster the rate of decolourisation, the higher the rate of the light-dependent stage.

Researching and referencing are important skills that are assessed by your teacher as evidence that you have met CPAC 5. You could take the opportunity to research relevant applications of your findings, such as:

- determining the most effective light intensity for photosynthesis in different plants
- describing the effects of different wavelengths of light on the rate of photosynthesis
- determining the optimum concentration of an inhibitor to use as a weedkiller
- selecting the most effective weedkiller

Focus on maths skills

Recording times

When recording the time taken, you should never include mixed units in a results table, i.e. minutes *and* seconds. Sometimes students record the reading of 6:30 (6 minutes and 30 seconds) from a stopwatch as 6.30 minutes when it should be 6.5 minutes (30 seconds is 0.5 minutes).

You should convert the minutes into seconds, so 6:30 would be recorded as 390 seconds.

Required practical 9

Factors affecting the rate of respiration in single-celled organisms

Background information

This practical requires knowledge of respiration, but also includes a synoptic link to enzymes, because respiration involves a series of enzyme-controlled reactions. Respiration produces ATP and has four main stages:

- glycolysis
- link reaction
- Krebs cycle
- oxidative phosphorylation

Glycolysis occurs in the cytoplasm and produces pyruvate, reduced NAD and ATP. The pyruvate is actively transported into the matrix of the mitochondria.

The link reaction occurs in the mitochondrial matrix and uses the pyruvate and coenzyme A to produce acetylcoenzyme A. This process also produces reduced NAD and carbon dioxide.

The Krebs cycle also occurs in the matrix and involves a series of oxidation–reduction reactions that produce reduced NAD, reduced FAD and ATP.

Oxidative phosphorylation takes place in the mitochondria using proteins, such as carriers and enzymes, embedded in the inner mitochondrial membrane. Electrons from reduced NAD and reduced FAD are transferred down the electron transfer chain (a series of electron carriers) and the energy released is used to actively transport protons across the inner mitochondrial membrane. The protons diffuse through ATP synthase down their concentration gradient and ATP is produced. The electrons and protons combine with oxygen and water is produced. Oxygen is the final electron acceptor.

> **Exam tip**
>
> Energy is *released* during respiration, not *produced*.

One of the ways of carrying out this practical involves using a dye such as methylene blue as an electron acceptor. The colour change can be seen easily and the time taken to change colour can be measured. The rate can be determined and used to investigate respiration because rate of colour change of the dye correlates to the rate of respiration.

As you will be investigating the effect of a factor on the rate of respiration, any factor that affects the enzymes involved in respiration or the proteins of the electron transfer chain will affect the overall rate of respiration. The factors that you could investigate are:

- temperature
- pH
- inhibitors

> **Exam tip**
>
> Glucose and other sugars are not the only respiratory substrates that can be used to produce ATP — both lipids and proteins are potential sources of energy.

Another factor that would affect the rate of respiration is the respiratory substrate used.

Guidance through the practical

One of the first decisions that you need to make is which factor you are going to change and how you are going to change it. Having made this decision, you need to determine a suitable range for your independent variable. Temperature is an obvious choice as it is easy to vary using a water bath, and can be easily monitored and controlled. Make sure that you choose a range that will encompass the organism's optimum temperature.

Once you have chosen the factor that you are going to investigate, you must ensure that other factors that could affect the rate of respiration are controlled. You should also set up a control experiment to show that any change is due to respiration in the organisms. An example might be setting up an identical experiment, but using a boiled culture of organisms.

This practical requires you to use cultures of single-celled organisms, so a suspension of yeast is a popular choice. Alternatively, your teacher may provide you with a culture of non-pathogenic bacteria.

There are two main approaches to this investigation:

- Using a respirometer. Simple respirometers can be used to determine the rate of oxygen uptake in aerobic respiration, and production of carbon dioxide in both aerobic and anaerobic respiration.

- Using a dye or stain that changes colour when reduced. Methylene blue is a blue dye that can act as an artificial electron acceptor and will decolourise when reduced. The time taken to decolourise can be measured and used to calculate the rate. The higher the rate of respiration, the faster the dye will decolourise.

Using a respirometer

As you can see from Figure 24, a simple respirometer has two key parts:

- An airtight tube or flask into which the living organisms can be placed together with a chemical, such as sodium hydroxide, that will absorb carbon dioxide.

- A capillary tube containing a drop of coloured liquid alongside a scale so that the distance travelled by the droplet can be measured.

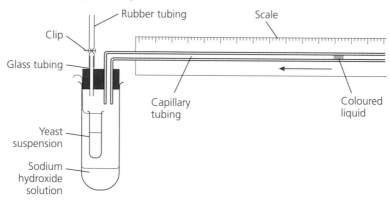

Figure 24 A simple respirometer used to measure the rate of oxygen uptake by a yeast culture

When organisms respire aerobically, oxygen is taken up and carbon dioxide is released. If living organisms are in a sealed container and the carbon dioxide is absorbed, the uptake of oxygen will reduce the pressure inside the container as the oxygen taken from the air is not being replaced by carbon dioxide. A respirometer works on the principle that as the pressure inside the sealed container drops, it becomes lower than atmospheric pressure, causing a drop of coloured liquid to move along the capillary tube. The faster the rate of respiration, the further the bubble will move in a set period of time.

The tube containing the organisms could be immersed in water baths at different temperatures to investigate the effect of temperature.

Using a dye as an electron acceptor

This method requires very little equipment and is a straightforward technique to carry out. The culture of single-celled organisms needs to be measured into a test tube along with glucose solution, then you need to measure the methylene blue into a second test tube. Both tubes should be left in the water bath at the required temperature for 5 minutes to **equilibrate**.

Equilibrate means 'to bring into equilibrium' and can be used in the context of a water bath to explain that a solution reaches the temperature of the water bath.

Variables that need to be controlled in this investigation are:
- volume of culture of single-celled organisms
- volume of glucose solution
- volume of methylene blue

Once the tubes have reached the required temperature, you should add the methylene blue to the tube containing the organisms and glucose, shake the tube to mix the contents, then replace the tube in the water bath and start the stopwatch. Record the time taken for the dye to decolourise (for the blue colour to disappear).

Determining when the solution has completely decolourised is *subjective*. You should have a culture of organisms without methylene blue as a comparison so that you can see when the blue colour has disappeared.

You can repeat this for each temperature so that a mean can be calculated. You should then repeat for at least four other temperatures.

Presentation of results from the practical

You will need to prepare a results table for your raw data before you start the practical. The factor that you are changing, such as temperature, will be in the first column, with units in the column heading only.

The subsequent columns in the table should be for your measurements of the distance moved by the drop of liquid. Depending on the apparatus, you may need columns for the start distance of the liquid, end distance of the liquid and the total distance moved. Alternatively, the respirometer that you use may allow you to reset the liquid to zero each time so that there would be no need to record the start and end positions of the bubble.

You will need a final column for processed data because this practical requires you to calculate *rate*. As rate is per unit time, you will need to record the distance travelled in a specific period of time, with the probable unit being $mm\,min^{-1}$. A better way of recording the rate would be to calculate the *volume* of oxygen taken up per minute ($mm^3\,min^{-1}$), as distance does not take the diameter of the capillary tube into account. See the maths skills box on page 48 for how to calculate this. An even better way of recording the data would be to calculate the rate of oxygen uptake per gram of organism ($mm^3\,min^{-1}\,g^{-1}$), in which case you would need to know the mass of yeast used in this practical (or other single-celled organism) used to make up the suspension.

If you have used the dye, you will still need to draw a table with the factor that you changed in the first column. The following columns will be for the time taken to decolourise in seconds (s), with enough columns for you to repeat each temperature and record the mean.

Process your mean time to give the rate for each temperature. This is calculated using 1/time, with a unit of s^{-1}.

Whichever method you follow, you can then plot a graph of your results. It will usually be a line graph because most of the factors you could investigate are continuous. (If you investigate different respiratory substrates, such as glucose, fructose and sucrose, you would draw a bar chart because the data are categoric.) The factor you investigate, for example temperature, will be on the *x*-axis and the rate on the *y*-axis. Draw a line of best fit, if appropriate, using the following rules:

Practical tip

Depending on your investigation, you could carry out a statistical test when analysing your results. A Student's *t*-test could be used to compare the means of two temperatures, or the correlation coefficient could be used to see if there is a relationship between temperature and rate.

- Ensure there are as many points on one side of the line as on the other.
- Draw a continuous line, using a sharp pencil.
- Do not **extrapolate** (extend the line beyond the first and last plotted points).
- Do not automatically ignore anomalous results.

Analysis/evaluation of the results

Once you have plotted your graph, you should describe the pattern. You then need to explain your findings using your scientific knowledge. This is a good opportunity to make synoptic links — for example, you may need to discuss the enzymes and other proteins involved in respiration, and explain how they have been affected by the factor that you investigated.

Practical tip

Remember that not all lines of best fit go through the origin. You need to consider whether an independent variable with a value of 0 would result in a value of 0 for the dependent variable. If so, then the line goes through the origin.

Focus on maths skills

Calculating surface areas and volumes

Calculating the surface area and volume of regular prisms, cylindrical prisms and spheres is one of the maths skills that may be assessed in the written exams. You need to learn the formulae shown in Figure 25 because they will not be provided in the exam.

For a regular prism:

surface area $= 2wl + 2lh + 2hw$

volume $= lwh$ ($l \times w \times h$)

where w = width, l = length, and h = height.

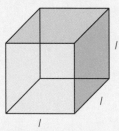

In the case of a regular cube, $l = w = h$ so:

surface area $= 6l^2$

volume $= l^3$

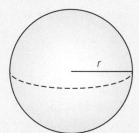

For a sphere:

surface area $= 4\pi r^2$

volume $= \dfrac{4}{3}\pi r^3$

where r = radius and the value of π is 3.14.

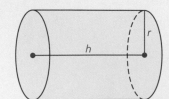

For a cylindrical prism:

surface area $= 2\pi rh + 2\pi r^2$

volume $= \pi r^2 h$

Figure 25 Formulae for calculating surface areas and volumes

Required practical 10

The effect of an environmental variable on animal movement

Background information

After studying the 'Stimuli and response' section of the specification, you will know that organisms respond to changes in their surroundings to increase their chances of survival. Motile organisms (organisms that can move) can show two types of simple response:

■ **Taxes (plural of taxis)** — these are simple directional responses that involve an organism moving towards a favourable stimulus or away from an unfavourable stimulus. Movements towards a stimulus are positive taxes and movements away from a stimulus are negative taxes. There are different types of taxes depending on the stimulus. For example, chemotaxis is a response to a chemical stimulus and phototaxis is a response to light.

■ **Kineses (plural of kinesis)** — these are non-directional responses that involve an organism changing its speed and turning frequency (Figure 26). When experiencing favourable conditions, an organism moves more slowly and turns more frequently so it remains in these favourable conditions. In unfavourable conditions, moving rapidly in a straight line increases an organism's chances of finding favourable conditions.

Exam tip

Students sometimes confuse taxes and kineses. One way of remembering is to think of a taxi — this will drive you straight to your destination rather than making frequent and random turns!

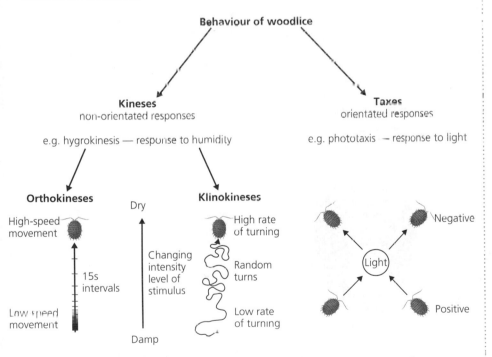

Figure 26 Types of behaviour shown by woodlice including **orthokinesis** (when the speed of movement is dependent on the intensity of the stimulus) and **klinokinesis** (when the frequency or rate of turning is proportional to the stimulus)

A choice chamber is a shallow perspex container that generally has a base divided into four and a lid with small holes through which organisms can be placed. Figure 27 shows the typical structure of a choice chamber. Using a choice chamber allows you to create up to four different environmental conditions then, after a period of time, count the number of organisms in each section to determine their preferred environment.

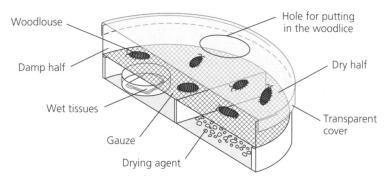

Figure 27 Cross-section through a choice chamber, showing how dry and damp conditions could be created

Although no animal is specified in the title of the investigation, invertebrates such as woodlice and maggots are often used in schools and colleges because they are readily available and show a preference for certain conditions.

Woodlice are land crustaceans that are 10–15 mm in length and feed on rotten wood and vegetation. They have external gills for gas exchange (Figure 28), which are covered with a thin film of moisture, so prefer cool, damp conditions and can be found easily under stones and logs.

Figure 28 Drawing of a woodlouse, showing the position of its gills

Maggots (Figure 29) are the larval stage of the European bluebottle and demonstrate negative phototaxis, behaviour that encourages them to burrow into their food (rotting food and faeces) and reduces their risk of predation and desiccation (drying out).

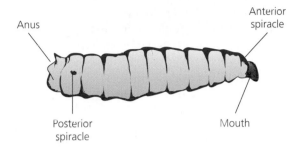

Figure 29 Drawing of a maggot, showing the main body parts

Guidance through the practical

Prior to completing this practical, a full risk assessment should be carried out either by you (as evidence for CPAC 3) or by your school or college. Handling animals can carry the risk of infection, and desiccants (drying agents) such as silica gel or anhydrous calcium chloride can be irritants. Referring to safety documents produced by CLEAPSS will inform you of appropriate control measures to take when handling these chemicals.

This is one of the few practicals that gives you the opportunity to handle living animals safely and ethically to measure their responses (ATh):

- Handling living animals safely will minimise the risk of infection associated with them. You should avoid eating and drinking in the lab, cover open wounds with a plaster and wash your hands thoroughly with soap and warm water before leaving the lab.
- Ethical handling of living animals refers to the way in which the animals are treated. Although they are simple organisms, they should still be treated with care and respect during the investigation.

If you have collected woodlice, then they should be returned to their habitat. Bluebottles are both a nuisance and carriers of disease, so the maggots should be killed humanely after use (by freezing for several days) and wrapped up securely before disposal.

When selecting the number of animals to use, bear in mind the number of conditions you are creating and the statistical test you intend to do. Statistical tests are discussed in more detail later in the practical. If you have four conditions, it makes sense to use a number of animals that is a multiple of four, to make subsequent calculations easier.

The practical title suggests the use of either a choice chamber or a maze. A choice chamber could be used to determine the preference of an organism for an environmental factor. You could investigate the preference for dry or humid conditions by adding a desiccant (dehydrating agent) such as silica gel or calcium chloride to one half of the chamber and damp filter paper or paper towel to the other half. You should leave the choice chamber to equilibrate for at least 5 minutes and check that the required conditions have been produced by testing with cobalt chloride strips, which are blue when dry and pink when moist. You should then insert the animals through the central hole, leave them for 5 minutes, then count and record the number of animals in each half. A preference for light or dark can also be investigated by covering one half with card, but you should ensure that the humidity and temperature are controlled.

Practical tip

The timings given for the investigation are for guidance only. You will need to adjust them depending on the animal you choose and its level of activity. Keep this time constant for any repeats.

You could also use a choice chamber to investigate kinesis by creating humid conditions in one half and dry conditions in the other. An animal could be inserted through a hole into the humid half and you could use a pen to trace its movement on the lid over a set period of time. Using this pen trace, you can measure the distance and count the number of turns made. If you repeat this process using a different animal each time and alternating humid and dry sections, you could then calculate the mean speed and the mean number of turns made in each condition.

You could make a simple maze out of paper or card, like the one shown in Figure 30, and use this to investigate taxes in animals. A food source could be placed at the end of either the left or right arm of the T-junction. You could then position an animal at position A and record whether it turns towards or away from the food source.

You should use a different animal each time and rub the inside of the maze with a cotton bud to remove any chemical trace left by the previous animal.

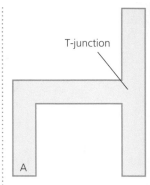

Figure 30 A simple maze that could be used to investigate animal behaviour

Presentation of results from the practical

A number of approaches have been described in the guidance above, but the presentation and analysis of results that follow will describe how to handle data from a choice chamber.

You should devise a suitable results table before starting the practical. This will probably be a simple table with the environmental condition in the first column and the number of animals present after 5 minutes in the second column.

A suitable number of repeats should be carried out and then the mean calculated and recorded in the final column.

Analysis/evaluation of the results

Chi-squared test

When analysing your results, you should use a statistical test to see if there is a significant difference in the distribution of the animals, or if any difference is just due to chance.

Step 1 Write a null hypothesis

You should start with a **null hypothesis**, which could be: 'There is no significant difference between the number of woodlice found in dark and light areas.'

Step 2 Choose a suitable statistical test and justify your choice

Use the flow chart given in Figure 31 (page 55) to help you choose a suitable test. You are collecting categoric data, so you would use the chi-squared (χ^2) statistical test. For this practical, you are comparing your observed values with the expected values. The expected value is the number of organisms you would expect to find in each condition if they showed no preference for a particular environment. If you have two conditions, such as light and dark, you would expect to find an equal number in each section.

A **null hypothesis** is a testable statement that there is no significant difference between the samples being tested.

Step 3 Calculate the test statistic

You then need to calculate chi-squared using the formula:

$$\chi^2 = \sum \frac{(O-E)^2}{E}$$

where O = your **observed** values (the number of animals that you counted in each section) and E = your **expected** values (an equal number of animals in each section). The easiest way to work through this calculation is by using a table, as shown in Table 7.

Table 7 How to calculate chi-squared — a worked example

Environmental condition	Observed number of animals (O)	Expected number of animals (E)	$(O-E)$	$(O-E)^2$	$\dfrac{(O-E)^2}{E}$
Dark	11	7	4	16	2.29
Light	3	7	−4	16	2.29
					$\Sigma = 4.58$

Step 4 Interpret the test statistic

Once you have calculated chi-squared, you need to look up the critical value of chi-squared in a table of probability. The number of degrees of freedom is your number of categories minus 1. The example above only has two categories, light and dark, so the number of degrees of freedom is 1. The critical value of chi-squared for 1 degree of freedom is 3.84 at the $p = 0.05$ level.

In the example above, the calculated value of chi-squared is 4.58, which is greater than the critical value of 3.84 at 1 degree of freedom.

If your value of chi-squared is *greater* than the critical value, then you *reject* the null hypothesis. There is *less* than 5% probability that the difference in distribution is due to chance. This means that there *is* a significant difference in the distribution of the animals, and you may be expected to suggest reasons for this difference.

If your value of chi-squared is *lower* than the critical value, then you *accept* the null hypothesis. There is *more* than a 5% probability that the difference in distribution is due to chance. This means that there is no significant difference in the distribution of the animals.

Focus on maths skills

Statistical tests

One of the maths skills that you may be assessed on is your ability to select and use a statistical test. There are three statistical tests in the specification: the chi-squared test, the Student's t-test and 'a test of correlation' (e.g. Spearman's rank correlation test).

The chi-squared test should be used if you are comparing frequencies of categoric data. This test is used to analyse the results of genetic crosses, because you can compare observed ratios with the ratios predicted from genetic diagrams. It is also useful for analysing data on animal behaviour, for example, whether a species of butterfly has a preference for a certain flower colour.

Exam tip

Lay out any calculations clearly and annotate fully. This may gain marks for your method, even if your final answer is incorrect.

Practical tip

Σ means 'the sum of', so you need to add the values in the final column.

Exam tip

There are different significance levels, but the $p = 0.05$ level is the one you will use. If the probability of the results being due to chance is less than or equal to $p = 0.05$ (5% or 1 in 20), then the difference is significant.

→

The Student's t-test is used when comparing the means of two sets of data. An example of when you might use this test is when comparing the mean diameter of limpets on the upper and lower sections of a rocky shore.

To calculate the value of t, you need to know:
- the means of each sample ($\bar{x}_1$ and $\bar{x}_2$)
- the number of measurements in each sample (n_1 and n_2)
- the variances of each sample (the variance is the square of the standard deviation)

A correlation coefficient is used when you are looking for associations between two sets of data. This is a useful statistical test when you are carrying out fieldwork investigations where, for example, you may be looking for a correlation between the distribution of a species and an abiotic factor. The correlation coefficient recommended by AQA is Spearman's rank, but others may be used.

Although you are expected to have used these statistical tests and to have a sound understanding of them, *you will not be asked to perform calculations using them in the exam papers*.

In the AS exams, you may be asked to *formulate* a null hypothesis for a practical that you have carried out or for a given experiment, and in the A-level exams you may be asked to *evaluate* a null hypothesis.

Your ability to *select* an appropriate statistical test and then *justify* this selection may be assessed in the AS exams, whereas the A-level exams may also ask you to *evaluate* another investigator's choice of statistical test. Figure 31 shows a flow chart that may help you choose an appropriate statistical test.

At AS you may be given a probability value and be expected to *interpret* whether the results are due to chance. At A-level you may be expected to *interpret* a probability value, with a statement about whether to accept or reject the null hypothesis using $p = 0.05$, as well as *evaluating* conclusions about data. At A-level you may also be given an extract from a table of probability (similar to the one shown in Table 8) and be expected to find a probability value for the appropriate number of degrees of freedom. For example, the critical value for t at the 5% ($p = 0.05$) level and with 2 degrees of freedom is 4.30.

Table 8 An extract from a statistical test table for values of t

Degrees of freedom	Significance level			
	10% ($p = 0.10$)	5% ($p = 0.05$)	1% ($p = 0.01$)	0.1% ($p = 0.001$)
1	6.31	12.71	63.66	636.62
2	2.92	4.30	9.93	31.60
3	2.35	3.18	5.84	12.94

→

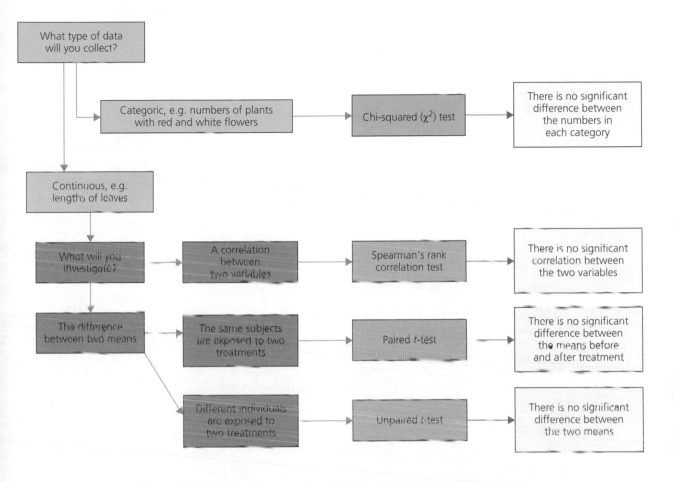

Figure 31 A flow chart to use when deciding which statistical test to use

Student's t-test

A student investigates the hypothesis that there is a difference between the diameter of limpets on the upper section of a rocky shore in Aberystwyth and those on the lower section of the shore. He measures the diameter of 12 limpets on the upper shore and 12 limpets on the lower shore and then works out the mean limpet diameter for each section of the shore. He then wants to carry out a statistical test to see if there is a significant difference between the mean diameters.

Step 1 Write a null hypothesis

A null hypothesis for this investigation could be: 'There is no significant difference between the mean diameters of limpets from the upper and lower shore.'

Step 2 Select a suitable statistical test and justify your choice

Referring to the flow chart in Figure 31, diameter is a continuous variable and the difference between two means is being compared. In this investigation, different groups of individuals are exposed to two treatments (the upper shore and the lower shore), so the student uses the unpaired t-test.

Step 3 Calculate the test statistic

The value of t is calculated using the formula:

$$t = \frac{\bar{x}_1 - \bar{x}_2}{\sqrt{\dfrac{s_1^2}{n_1} + \dfrac{s_2^2}{n_2}}}$$

where:

- $\bar{x}_1$ and $\bar{x}_2$ are the means for the upper and lower shore
- s_1^2 and s_2^2 are the variances for the upper and lower shore
- n_1 and n_2 are the number of limpets measured on the upper and lower shores

The results from the investigation are shown in Table 9.

Table 9 Data on limpet diameter from the upper and lower shores

	Upper shore	Lower shore
Number of limpets measured (n)	12	12
Mean diameter/mm	13.2	7.9
Standard deviation (s)/mm	2.17	1.16
Variance (s^2)	4.71 (to 2 d.p.)	1.35 (to 2 d.p.)

The figures from Table 9 can be substituted into the formula to give a value for t:

$$t = \frac{13.2 - 7.9}{\sqrt{\left(\dfrac{4.71}{12} + \dfrac{1.35}{12}\right)}}$$

$$t = \frac{5.3}{\sqrt{0.505}}$$

$$t = \frac{5.3}{0.7106}$$

$$t = 7.458 \text{ (to 3 d.p.)}$$

Step 4 Interpret the test statistic

Once the value of t has been calculated, the critical value of t is found for $(n_1 + n_2) - 2$ degrees of freedom. As 12 measurements were taken at each site, the number of degrees of freedom for the worked example is 22. Table 10 shows an extract from a statistical table for values of t.

Table 10 Critical values of t

Number of degrees of freedom	Significance level			
	$p = 0.10$ (10%)	$p = 0.05$ (5%)	$p = 0.01$ (1%)	$p = 0.001$ (0.1%)
22	1.717	2.074	2.819	3.792

Practical tip

Although standard deviation can be calculated manually, it is expected that you will use a calculator. If you are not sure how to do this, find someone with the same type of calculator as you and ask them to show you how to do it, or check in your instruction manual.

The calculated value of t of 7.458 is much greater than the critical value of 3.792 at $p = 0.001$ (0.1%). This means that the probability of the difference between the means occurring by chance is less than 0.1%, so it is therefore *highly unlikely* that the difference between the two means could have occurred by chance. The *null hypothesis is rejected* and the difference between the mean diameters of the limpets is significant.

Required practical 11

Using a calibration curve to identify an unknown concentration

Background information

This practical is linked to the specification section 'Control of blood glucose concentration', which requires you to know the roles of the liver and hormones, including insulin, glucagon and adrenaline.

Exam tip

There are many similar-sounding words in this section that students sometimes get mixed up. Glucagon is a hormone and glycogen is a storage polysaccharide. You also need to learn the difference between glycogenesis, gluconeogenesis and glycogenolysis.

You should also know the causes and control of type I and type II diabetes. There are a number of **signs** and **symptoms** of diabetes, with glucose in the urine being one of the signs.

Normally, the urine contains no glucose because it is all reabsorbed in the proximal convoluted tubule, but high levels of glucose in the blood can lead to glycosuria (glucose in the urine). Glycosuria leads to increased water loss in the urine, so frequent urination is a symptom of diabetes. The presence of glucose in urine can be shown using reagent test strips, such as Diastix®, which are dipped into urine and change colour to indicate the concentration of glucose present. In this practical you will use Benedict's reagent to test a solution made up by the technician to resemble urine. This required practical does *not* require you to test real urine!

The Benedict's test is a test for reducing sugars, which includes all monosaccharides and some disaccharides. Sucrose is a disaccharide and is not a reducing sugar. Benedict's solution contains copper(II) sulfate, which is reduced to copper(I) oxide when heated with a reducing sugar. This forms an insoluble red precipitate and so the presence of a reducing sugar can be seen by a colour change.

The Benedict's test can be used **quantitatively** because the mass of precipitate formed correlates to the concentration of reducing sugar present. The test can also be used **semi-quantitatively** because the colour obtained gives an indication of the concentration of reducing sugar present. The solution remains blue in the absence of reducing sugar, then changes colour to green, yellow, orange or brick-red with increasing reducing sugar concentration, as illustrated in Figure 32.

Glycogenesis is the conversion of glucose to glycogen. **Gluconeogenesis** is the production of glucose from non-carbohydrate sources such as amino acids. **Glycogenolysis** is the breakdown of glycogen to glucose.

Signs of a disease can be seen or measured, whereas **symptoms** are felt by the patient. A high temperature and rash are signs, whereas pain and tiredness are symptoms.

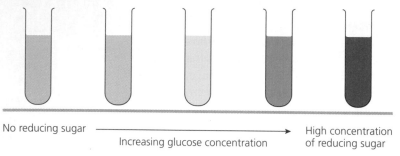

No reducing sugar ⟶ Increasing glucose concentration ⟶ High concentration of reducing sugar

Figure 32 Diagram showing the approximate colours of Benedict's reagent with a range of different glucose concentrations

Guidance through the practical

This is a straightforward practical and two of the key skills — using glassware for dilutions (ATc) and using a colorimeter to record quantitative measurements (ATb) — have already been covered in detail in required practicals 3 and 4. Additionally, you will use Benedict's solution to identify the presence of a reducing sugar (ATf).

The first stage of this practical is preparing your dilution series using a stock glucose solution and distilled water. The stock solution and the distilled water will be two of your test samples, and you should prepare at least four intermediate concentrations. If you were given a stock solution of $10.0\,\text{mmol}\,\text{dm}^{-3}$ concentration, you could use this to prepare intermediate concentrations of 2.0, 4.0, 6.0 and $8.0\,\text{mmol}\,\text{dm}^{-3}$.

Once you have made your dilutions, you need to carry out the Benedict's test on each sample. This involves adding Benedict's solution to your test samples and heating them in a water bath of hot water (about 70°C). There are a number of variables that you need to control to allow comparison of the final colours:

- volume of each test sample
- volume of Benedict's solution
- temperature of water bath
- time in the water bath

You should now have at least six tubes of known glucose concentration that have been heated with Benedict's solution and then left to cool. You should also have at least one unknown sample of fake urine (prepared by your teacher or technician) that has also been heated with Benedict's solution.

The second stage of the practical is taking the absorbance readings, using a colorimeter.

You can remove the precipitate by filtering each solution into a clean test tube labelled with the glucose concentration. Depending on the colorimeter that you are using, you may be able to leave the solutions in the test tubes, or you may need to transfer the contents of each tube to a clean cuvette.

When setting up the colorimeter, you will need to use the contents of the tube with $0.0\,\text{mmol}\,\text{dm}^{-3}$ glucose tested with Benedict's solution as your blank, not just distilled water. Record the absorbance of each solution in your results table.

Exam tip

If asked to describe the test for a reducing sugar in the exam, remember that you add Benedict's solution and *heat* in a hot water bath.

Presentation of results from the practical

Draw a results table with the concentration of glucose solution/mmol dm^{-3} in the first column and absorbance in the second column. Remember that there are no units for absorbance because it is a ratio.

You then need to plot a graph of your results, with concentration of glucose solution on the x-axis and absorbance on the y-axis. Draw a suitable line of best fit through your points. You may find that it is a straight line, or it may be a curve, depending on the concentrations of glucose used. The graph is still called a **calibration curve** even if it is a straight line.

In this practical, if you *have not filtered* the solutions, you would expect absorbance to *increase* (or percentage light transmission to *decrease*) as the concentration of glucose *increases*. This is because at a high glucose concentration a greater mass of precipitate would form due to the reaction between the glucose and the copper sulfate ions, and more light would be absorbed by the solution.

If you *have filtered* the solutions to remove the precipitate, then absorbance would *decrease* as the concentration of glucose *increases*. At low concentrations of glucose, few Cu(II) ions are reduced to Cu(I) so the solution remains a darker blue and gives a higher absorbance. More reduction of Cu(II) at higher glucose concentrations produces a lighter blue solution and so a lower absorbance.

Analysis/evaluation of the results

Once you have plotted your calibration curve, you can use it to find the glucose concentration of your unknown solution. Find the absorbance of the unknown solution on the y-axis and read across until you reach the curve; then read down to find the value of the intercept on the x-axis.

Figure 33 shows that an unknown solution has an absorbance of 1.00. Reading across to the curve and then down to the x-intercept, the glucose concentration of the unknown solution is found to be 3.7%.

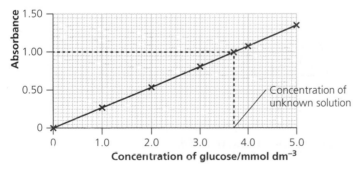

Figure 33 How to use a calibration curve to determine the x-axis intercept

Practical tip

When you are using a calibration curve to determine an unknown value, use a ruler to draw dashed lines from the measured value on the y-axis to the curve, then from the curve down to the intercept on the x-axis.

Focus on maths skills

Representing uncertainty

When plotting a point on a graph, there is always a level of uncertainty. You may be plotting a mean value, in which case your level of uncertainty depends on the values used to calculate the mean, and the presence or absence of anomalous results.

Range bars are the simplest way of showing the spread of data around the mean. The maximum and minimum values used to calculate the mean are marked above and below the plotted point using small lines, and the two lines are joined by a vertical line through the plotted point.

Error bars are drawn in exactly the same way as range bars, but may be shown as plus and minus one standard deviation above and below the plotted point (or bar), as shown in Figure 34. Using the standard deviation rather than the range is preferable because calculating standard deviation still shows the spread of data around the mean, but minimises the effect of extreme values. The longer the error bars, the greater the spread of data around the mean.

If you draw a line of best fit that does not go through each plotted point, then it should be drawn within the error bars.

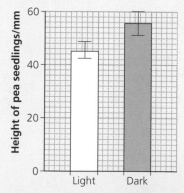

Figure 34 A graph showing the height of pea seedlings grown in the light and in the dark. Error bars have been included to show the spread of data around the mean

Required practical 12

The effect of environmental factors on species distribution

Background information

The location chosen for this practical activity will depend on the amount of time available and the proximity of your school or college to a suitable site. Some schools and colleges visit specialist fieldwork centres, with popular choices for this type of investigation being sand dunes and rocky shores.

The study of the interactions between organisms and their environment is called **ecology**. The environment includes both abiotic (non-living) and biotic (living) factors. Abiotic factors include:

- light intensity
- temperature
- oxygen concentration
- pH

Biotic factors include:

- predation
- competition
- disease

You will have studied the effects of these abiotic and biotic factors on populations in the 'Populations in ecosystems' section of the specification.

Your choice of sampling technique will depend on the factor you are investigating and the hypothesis you have developed, but the two main techniques are random sampling and systematic sampling.

Random sampling

This is a technique designed to remove **sampling bias**. The person carrying out the investigation may deliberately or unconsciously make a biased choice, for example putting a quadrat down on a patch of grass with the most daisies. One random sampling technique is to divide the area being investigated into a grid using two tape measures as axes. Random numbers can be generated using a table of random numbers, or a random number generator on a calculator. These random numbers can be used as coordinates, with a sample being taken where the coordinates intersect, as shown in Figure 35. Random sampling does not involve throwing a quadrat over your shoulder without looking!

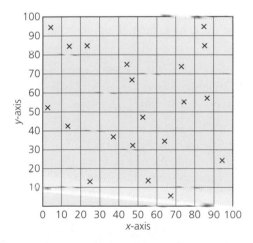

Figure 35 How to place quadrats using random number coordinates

Systematic sampling

This technique involves taking samples at regular intervals and is useful if you are investigating a correlation, or if you are looking at the stages of succession. An example of a systematic sampling technique is an interrupted belt transect (Figure 36). A tape measure could be laid across the ground and then samples taken at regular intervals, for example every 1 m or every 10 m (depending on the total distance to be covered).

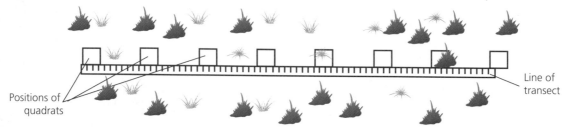

Positions of quadrats

Line of transect

Figure 36 How to carry out an interrupted belt transect

Both techniques described above may require the use of a quadrat (Figure 37). This is a square frame made of metal or plastic, usually $0.25\,m^2$, that is divided into smaller squares and can be used to sample plants or non-motile organisms. The number of organisms could be counted, or the percentage cover could be estimated.

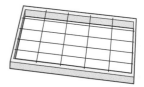

Figure 37 A typical wooden-framed quadrat

Guidance through the practical

This is an open-ended investigation, which lends itself to a full planning activity that could provide evidence for all of the CPAC. Unlike some of the required practicals, very little equipment is necessary and the investigation could be carried out completely independently in your garden or a nearby park.

During this practical, the main skill you will develop is the use of sampling techniques in fieldwork (AT k). The sampling technique you choose will depend on your hypothesis, and on your choice of both environmental factor and organism. You need to decide if random sampling or systematic sampling is more appropriate for your investigation.

Once you have decided on a sampling technique, you need to choose a sample size. You need a suitable number of samples for statistical analysis. A large sample size will also minimise the effect of anomalies and give more representative measurements.

The size of the species being investigated and the size of the area being investigated will affect the *size of quadrat* that you choose. The area being sampled and the available time will determine the *number of quadrats* that you include in your sample, but you could carry out some preliminary work to determine the optimum quadrat number to use.

As you are investigating the effect of an environmental factor, you also need a method of measuring this factor. For example, a light meter could be used to measure light intensity, the pH of soil or water could be measured using a pH probe, and nitrate test strips could be used to determine the concentration of soil nitrates.

Some possible investigation titles include:

■ The effect of wind speed on the distribution of marram grass on a sand dune
■ The effect of light intensity on the distribution of daisies on a lawn
■ The effect of oxygen concentration on invertebrate distribution in a river

Presentation of results from the practical

Preparing a results table prior to this investigation is important, because you will be outside collecting the data. It may be a good idea to prepare a table for your raw data on a piece of paper attached to a clipboard rather than risking getting your lab book wet or dirty. Weatherproof paper or a plastic wallet will allow you to keep a clear record of your data in adverse weather conditions.

Different types of graph may be drawn; you should select the one most appropriate for your investigation.

You may choose to draw a bar chart if the independent variable is categoric. As the independent variable is not continuous, the bars should not touch.

Analysis/evaluation of the results

You will need to carry out a statistical analysis of your results to find out if your results are significant. If you find that your results are significant, then you should research the reasons for them and explain your observations using your biological knowledge. If your results are not statistically significant, you should explore possible reasons for this, with an evaluation of your methodology and possible improvements to your investigation.

This analysis provides an excellent opportunity for you to make synoptic links:

■ If you have investigated temperature, you could discuss the effect on enzymes.
■ For light intensity, you could make links to photosynthesis.
■ Salt concentration could be linked to osmosis and xerophytic adaptations.
■ Nitrate concentration could be discussed with reference to the nitrogen cycle, protein synthesis and growth.

Focus on maths skills

Correlation

Scatter diagrams or scattergrams are used to identify a correlation between two variables (Figure 38) and you may be assessed on your ability to interpret them. The relationship between the two variables could be:

■ no correlation — there is no relationship between the two variables
■ positive correlation — as one variable increases, the other variable increases
■ negative correlation — as one variable increases, the other variable decreases

→

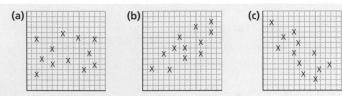

Figure 38 Scattergrams showing three possible trends: (a) no correlation; (b) positive correlation; (c) negative correlation

Just because there is a **correlation** between two variables, it does not mean that there is a **causal** link. There may be a correlation between ice cream sales and cases of hay fever, but that does not mean that ice cream causes hay fever.

The correlation coefficient

When you plan an investigation looking for a correlation or association between two variables, you start with a **hypothesis**, a testable statement. For example:

- There is a correlation between the distance from the strand line on a rocky shore and the number of bladders present on bladder wrack (a type of seaweed).
- There is a correlation between the distance from the strand line and the concentration of nitrate in the soil on the sand dunes.
- There is a correlation between the percentage cover of marram grass on the sand dunes and the wind speed.
- There is a relationship between the distance from the upper section of a rocky shore and the diameter of limpets.

Once you have completed your investigation and collected your data, you need to carry out a statistical test to determine whether or not the correlation is significant.

A student hypothesises that there is a correlation between wind speed and the percentage cover of marram grass on the sand dunes in Aberystwyth. At 12 randomly selected points in the dunes, she uses a quadrat to estimate the percentage cover of marram grass and an anemometer to measure the wind speed. The following steps describe how to analyse the data that she collects.

Step 1 Write a null hypothesis

The **null hypothesis** states that there is no relationship or correlation. For example: 'There is no correlation between the percentage cover of marram grass on the sand dunes and wind speed.'

Step 2 Select a statistical test and justify your choice

Choose a suitable statistical test using the flow chart in Figure 31 (page 55). As you are looking for a correlation between two variables, you will use the **Spearman's rank** correlation test. This test shows the strength and direction (positive or negative) of a relationship between two variables.

Step 3 Calculate the test statistic

As you are using Spearman's rank correlation test, you need to calculate the correlation coefficient using the formula:

$$r_s = 1 - \frac{6\sum D^2}{n^3 - n}$$

Exam tip

Do not write casual instead of **causal**.

Practical tip

Spearman's rank is one type of correlation test, but other statistical tests can be used, for example Pearson's product moment correlation test. In each case you calculate a correlation coefficient.

where:

- r_s = correlation coefficient
- Σ = sum of
- D = difference between the ranks
- n = number of pairs of measurements

Table 11 shows a straightforward method of presenting your results when you are calculating the Spearman's rank correlation coefficient.

Table 11 How to calculate the Spearman's rank correlation coefficient — a worked example

Wind speed/m s^{-1}	Rank wind speed	Percentage cover of marram grass	Rank percentage cover	Difference between ranks (D)	D^2
1.4	4	80	4	0 (4 – 4)	0
1.2	6	100	1	5 (6 – 1)	25
0.2	12	60	7.5	4.5 (12 – 7.5)	20.25
1.3	5	70	6	–1	1
0.3	10.5	40	9	1.5	2.25
0.3	10.5	20	11	–0.5	0.25
1.9	2	80	4	–2	4
0.6	8	30	10	–2	4
2.8	1	80	4	–3	9
1.8	3	90	2	1	1
0.7	7	60	7.5	–0.5	0.25
0.4	9	0	12	–3	9
					ΣD^2 = 76

Practical tip

When you are ranking values, write a list of numbers and cross out each number as you use it in your ranking so that you do not accidentally use the same rank too many times. This is particularly useful if you have values that are the same.

1 Rank the wind speed by giving the highest value a '1', the second highest a '2' and so on. If you have two values the same, like 0.3 m s^{-1} in Table 11, they 'share' ranks 10 and 11 and are both given a rank of 10.5. The next value in Table 11, 0.2 m s^{-1}, then takes rank 12.

2 Rank the percentage cover of marram grass, again giving the highest percentage cover a rank of '1'. If you have three values the same, like 80% in Table 11, they share ranks 3, 4 and 5 and are all given ranks of 4 (the mean of 3, 4 and 5). The next value, 70%, takes rank 6.

3 Find the difference between the ranks (D) by subtracting each rank for percentage marram grass cover from the rank for wind speed.

4 Calculate D^2.

5 Add up all of the values of D^2 (ΣD^2) = 76

6 Multiply ΣD^2 by 6 ($6\Sigma D^2$) – 6 × 76 = 456

7 Calculate $n^3 - n = 12^3 - 12 = 1728 - 12 = 1716$

8 Find $6\Sigma D^2/(n^3 - n) = 456/1716 = 0.2657342$

9 Find $1 - [6\Sigma D^2/(n^3 - n)] = 0.7342657 = 0.734$ (to 3 d.p.)

Don't forget the last step. This is commonly overlooked by students.

You have now calculated the test statistic and have a value of 0.734 for the correlation coefficient. Your answer should be given to three decimal places because critical values in probability tables are usually given to three decimal places.

Step 4 Interpreting the test statistic

Your calculated value of r_s will be between –1 (a perfect negative correlation) and +1 (a perfect positive correlation). The closer your calculated value is to 0, the weaker the correlation.

Once you have calculated the test statistic, you need to look up the critical value of Spearman's rank for the number of pairs of measurements (n). Table 12 shows an extract from a probability table.

Table 12 Critical values for Spearman's rank correlation coefficient

Number of pairs of measurements (n)	Significance level			
	$p = 0.10$ (10%)	$p = 0.05$ (5%)	$p = 0.01$ (1%)	$p = 0.001$ (0.1%)
10	0.564	0.648	0.794	0.903
12	0.503	0.587	0.727	0.846
14	0.464	0.538	0.679	0.802

There were 12 pairs of measurements in the worked example, so the critical value for r_s for 12 pairs of measurements at the 5% level ($p = 0.05$) is 0.587.

The calculated value of 0.734 is *greater than* the critical value of 0.587 at the 5% level, so the *null hypothesis is rejected*. There *is a correlation* between wind speed and the percentage cover of marram grass.

The value of 0.734 falls between the critical values for $p = 0.01$ and $p = 0.001$ for 12 pairs of measurements, so there is a probability of between $p = 0.01$ (1%) and $p = 0.001$ (0.1%) that this result occurred by chance.

As the calculated value is a positive number, it indicates a positive correlation, so as wind speed increases the percentage cover of marram grass also increases.

If you have established that there is a correlation, you can investigate whether or not there is a **causal relationship**. Just because two variables are linked, this does not prove that one causes the other. There could be a third variable that has not been investigated.

Questions & Answers

The exams

At AS there are two 1½-hour examination papers, each worth 50% of the total AS grade. Papers 1 and 2 assess any of the content from topics 1–4, including practical skills. At A-level there are three 2-hour exam papers. Paper 1 assesses any of the content from topics 1–4, paper 2 assesses the content from topics 5–8 and paper 3 assesses any content from the specification. Papers 1 and 2 are worth 35% each and paper 3 is worth 30%. Paper 3 includes questions requiring you to critically analyse experimental data. It also includes a 25-mark essay from a choice of two titles.

About this section

This section contains questions on some of the required practicals. These may appear in any of the papers in both the AS and the A-level exams. They are written in the same style as the questions in the exam so they will give you an indication of what you can expect. After each question there are answers by two different students.

Comments on the questions are preceded by the icon ⓔ. They offer tips on what you need to do to gain full marks. All student responses are followed by examiner's comments, indicated by the icon ⓔ, which highlight where credit is due. In the weaker answers, they also point out areas for improvement, specific problems and common errors such as lack of clarity, irrelevance, misinterpretation of the question and mistaken meanings of terms.

Question 1 The effect of pH on starch hydrolysis

A student investigated the effect of pH on the time taken by amylase to hydrolyse starch. The student:

1 added amylase and starch suspension to a test tube containing a buffer solution at pH 3

2 stood the test tube in a water bath and started a stopwatch

3 took samples at regular intervals and tested them with iodine solution

4 continued to take samples until a blue-black colour no longer appeared, and then recorded the time

5 repeated steps 1–4 using buffers at pH 5, 7, 9 and 11

Questions & Answers

(a) The student controlled the temperature. State *two* other variables that she should have controlled. (2 marks)

ⓔ This question just requires you to state two variables, but these should be *other than* temperature, which is given in the question.

(b) Describe how the student could have monitored the temperature. (2 marks)

ⓔ Note that there are 2 marks for this question. Which piece of equipment would you use *and* how would you use it?

(c) The student maintained the temperature at 30°C. Explain why the rate of starch hydrolysis would be faster at 40°C. (2 marks)

ⓔ The command word is *explain*, so this question requires you to use your knowledge of collision theory from topic 1: Biological molecules.

(d) The student took samples until the blue-black colour no longer appeared. How could she have ensured that her decision was consistent? (1 mark)

ⓔ You need to describe what you could compare each sample against when determining the end-point.

The student calculated the rate of reaction and plotted a graph of the results, as shown in Figure 1.

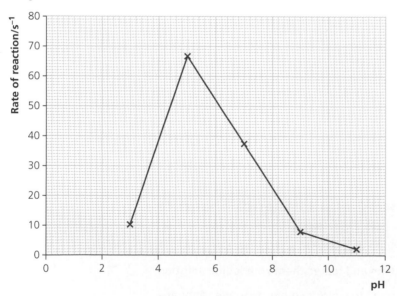

Figure 1 Graph to show the effect of pH on the rate of starch hydrolysis by amylase

(e) Explain why she joined the points with straight lines rather than drawing a line of best fit. (1 mark)

ⓔ Look at the intervals between each pH that the student tested.

(f) The student concluded that pH 5 was the optimum pH for this enzyme because the results show that the rate was highest at pH 5. Explain why this conclusion may not be valid. (2 marks)

(e) The student has only found pH 5 to have the fastest rate of the pHs that she tested. Think about a further investigation that would give an optimum pH closer to the true value.

Student A

(a) concentration of amylase and temperature

(e) **1/2 marks awarded** Temperature does need to be controlled, but this variable is given in the question.

(b) She could stand a thermometer in the water bath and check the temperature every 10 minutes to make sure it stays constant.

(e) **2/2 marks awarded** Student A clearly describes a method that could be used to monitor the temperature. The answer does not need to specify 10-minute intervals as long as the temperature is checked regularly.

(c) The starch and amylase have more kinetic energy at higher temperatures so there are more successful collisions and more enzyme–substrate complexes form.

(e) **2/2 marks awarded** This answer clearly explains the effect of a higher temperature on the enzyme and substrate, and why this would increase rate.

(d) She could have compared the colour of each sample to a sample of reaction mixture that has gone to completion and been tested with iodine solution.

(e) **1/1 mark awarded** This is an excellent answer that would allow the determination of a consistent end-point.

(e) She cannot be sure of the intermediate values, so she should join each point with a straight, ruled line.

(e) **1/1 mark awarded** This is an excellent answer.

(f) The student needs to repeat the investigation using pH 4, 6, 8 and 10.

(e) **1/2 marks awarded** Student A has realised that intermediate values need to be tested, but the optimum cannot be above 7 as the rate is decreasing. A better answer would suggest testing around pH 5, for example pH 4, 4.5, 5.5 and 6

Student B

(a) volume of amylase and pH

ⓔ 1/2 marks awarded Although pH is a factor that affects the rate of an enzyme-controlled reaction, in this investigation pH is the independent variable.

(b) She could check the temperature using a thermometer.

ⓔ 1/2 marks awarded There needs to be the idea of taking the temperature at regular intervals to gain the second marking point.

(c) The enzyme and substrate molecules move around faster and so the rate is faster.

ⓔ 0/2 marks awarded This answer lacks scientific terminology — there is no reference to 'kinetic energy', 'collisions' or 'enzyme–substrate complexes'. It also repeats the stem of the question rather than explaining *why* the rate is faster.

(d) The student could mix amylase with water and test with iodine solution, and then compare her samples with this colour.

ⓔ 1/1 mark awarded This is a good suggestion that would also allow a consistent end-point to be determined.

(e) The student should draw a curve.

ⓔ 0/1 mark awarded Student B is confused because many graphs show the effect of pH on the rate of an enzyme-controlled reaction as a curve. This does not answer the question.

(f) The student needs to repeat the investigation using a bigger range of pH.

ⓔ 0/2 marks awarded A bigger range suggests using pH below 3 and above 11, which would not provide a more accurate value for the optimum pH.

Question 2 Preparing an onion root tip squash

A student removed the tip from an onion root and put it into $5\,mol\,dm^{-3}$ solution of hydrochloric acid to kill the cells. He then followed instructions to separate and stain the cells. He placed a coverslip on top of the root tip and pressed down firmly, then viewed the slide using an optical microscope.

Figure 2 shows some of the cells he viewed.

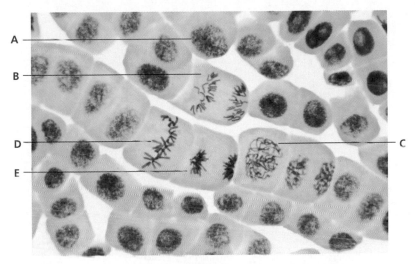

Figure 2 Cells from the tip of a plant root as seen using an optical microscope

(a) Cell A is undergoing interphase. State what happens to the DNA during interphase. (1 mark)

🄮 Remember that interphase is a stage of the cell cycle and *not* a stage of mitosis.

(b) Name the stages of mitosis shown by cells B, C, D and E. (4 marks)

🄮 As well as knowing what happens at each stage of mitosis, you need to familiarise yourself with diagrams and photos of each stage.

(c) Explain why the student added a drop of stain to the slide. (1 mark)

🄮 There are many different stains used in the preparation of slides, but they all have the same general function.

(d) Explain why the coverslip was pressed down firmly. (1 mark)

🄮 A root tip would be a piece of tissue consisting of many layers of cells. What would you need to do if you wanted to view individual cells?

(e) Suggest why the student handled the hydrochloric acid with caution. (1 mark)

🄮 Look at the concentration of the hydrochloric acid. How does this compare with the concentrations you may have used during your other practical activities? The unit mol dm^{-3} (moles per dm^3) is also written as 'M'.

Student A

(a) The DNA is replicated.

🄮 **1/1 mark awarded** Correct — the mass of DNA doubles during interphase, which takes up most of the cell cycle, and is followed by nuclear division and cytokinesis.

(b) B — anaphase, C — prophase, D — metaphase, E — telophase

🄮 **4/4 marks awarded** Student A has correctly recognised the stages of mitosis.

Practical assessment 71

(c) The stain binds to the chromosomes, so they become visible.

ⓔ **1/1 mark awarded** Correct.

(d) This was to squash the root tip into a thin layer so that light could pass through the sample.

ⓔ **1/1 mark awarded** This answer makes an important point — an optical microscope depends on the specimen being thin enough to allow the transmission of light.

(e) $5\,mol\,dm^{-3}$ acid is highly concentrated and would be corrosive.

ⓔ **1/1 mark awarded** Student A has recognised that this concentration is very high and has correctly used the term 'corrosive'.

Student B

(a) Nothing happens to the DNA during interphase. The cell is resting between divisions.

ⓔ 0/1 mark awarded The idea that the cell is 'resting' during interphase is a common misconception. Although the cell is not dividing during interphase, the DNA is being replicated.

(b) B — metaphase, C — prophase, D — anaphase, E — telophase

ⓔ 2/4 marks awarded Student B has mixed up anaphase and metaphase. Remember that the chromosomes line up in the **M**iddle of the cell during **M**etaphase and move **A**part during **A**naphase.

(c) So he could see the chromosomes.

ⓔ 1/1 mark awarded This answer is not as detailed as student A's answer, but still gains the mark.

(d) To flatten the cells.

ⓔ 0/1 mark awarded Pressing down on the coverslip would not squash individual cells, but would spread the cells out so they were in a single layer.

(e) The acid is very concentrated and could cause burns.

ⓔ 1/1 mark awarded Although student B has not used the term 'corrosive', they have clearly suggested why the acid should be handled with caution.

Question 3 The effect of sucrose concentration on osmosis

A student labelled six boiling tubes 0.0, 0.2, 0.4, 0.6, 0.8 and 1.0 mol dm^{-3} sucrose solution. She used distilled water and a 1.0 mol dm^{-3} sucrose solution to make up 20 cm^3 of each sucrose concentration in the boiling tubes. She then:

1   removed the 'skin' from a potato

2   cut six potato chips to the same length

3   weighed the chips and recorded their masses

4   put a chip into each boiling tube and left them for 30 minutes

5   gently blotted the potato chips dry

6   reweighed them and recorded their masses

7   repeated twice more to obtain three results for each concentration

She processed her results and found the percentage change in mass of the chip for each concentration. The student's results are shown in Table 1.

Table 1 The effect of sucrose concentration on the mass of potato tissue

Concentration of sucrose solution/ mol dm^{-3}	Initial mass/g	Final mass/g	Change in mass/g	Percentage change in mass	Mean percentage change in mass
0.0	1.26	1.49	0.23	18.3	19.3
	1.26	1.51	0.25	19.8	
	1.22	1.46	0.24	19.7	
0.2	1.22	1.31	0.09	7.4	5.4
	1.27	1.32	0.05	3.9	
	1.25	1.31	0.06	4.8	
0.4	1.43	1.29	−0.14	−9.8	−7.7
	1.32	1.26	−0.06	−4.5	
	1.37	1.25	−0.12	−8.8	
0.6	1.32	1.10	−0.22	−16.7	−15.2
	1.26	1.07	−0.19	−15.1	
	1.22	1.05	−0.17	−13.9	
0.8	1.28	0.98	−0.30	−23.4	−21.2
	1.25	1.01	−0.24	−19.2	
	1.23	0.97	−0.26	−21.1	
1.0	1.31	0.96	−0.35	−26.7	−23.3
	1.27	1.00	−0.27	−21.3	
	1.23	0.96	−0.27	−22.0	

(a) Why did the student use potato chips from the same potato?   (1 mark)

ⓔ Think of possible differences between cells from different potatoes.

(b) Describe how the student made $20\,cm^3$ of $0.6\,mol\,dm^{-3}$ sucrose solution. (1 mark)

ⓔ Although there is only 1 mark for the question, you need to state the volume of $1.0\,mol\,dm^{-3}$ sucrose solution used *and* the volume of distilled water.

(c) Suggest why the student removed the 'skin' from the potato. (1 mark)

ⓔ This is a 'suggest' question, which means you have to give a possible reason. In this question, you should consider how the 'skin', or the cells of the 'skin', might affect the investigation.

(d) Explain why the student processed her data to find the percentage change in mass. (1 mark)

ⓔ The key idea behind this question is that the potato cylinders did not have the same initial mass.

(e) Explain why the potato chip left in distilled water gained mass. (2 marks)

ⓔ Your answer needs to explain the direction of water movement in terms of osmosis and water potential.

(f) Describe how the student could use her processed data to find the water potential of the potato tissue. (3 marks)

ⓔ To answer this question, you need to describe how she would plot a graph, how she would determine the sucrose concentration of the potato tissue, and how she could then find the water potential.

Student A

(a) If the chips are from the same potato, all of the cells will have the same water potential.

ⓔ **1/1 mark awarded** Using the same potato means that the whole potato has been exposed to the same conditions before being cut into pieces.

(b) She used a measuring cylinder to measure out $12\,cm^3$ of $1.0\,mol\,dm^{-3}$ sucrose solution and poured it into a boiling tube. She then used a fresh measuring cylinder to measure $8\,cm^3$ of distilled water and poured this into the same boiling tube.

ⓔ **1/1 mark awarded** This is a detailed answer — only the volumes of solutions are required for the mark.

(c) The 'skin' is a different tissue from the rest of the potato and so would have a different water potential.

ⓔ **1/1 mark awarded** Student A gains the mark for stating that the 'skin' is a different tissue and for recognising that different tissues may have different water potentials.

(d) The initial mass of the potato chips was different, so working out the percentage change would allow her to make a valid comparison between the chips.

ⓔ **1/1 mark awarded** A correct and clear answer.

(e) The potato chips gained mass because water moved into the cells by osmosis. The water moved from a less negative ψ outside the cells to a more negative ψ inside the cells.

ⓔ **2/2 marks awarded** Student A has described the direction of water movement, stating that this was by osmosis, and correctly described the water potential as more negative inside the cells. The symbol ψ, representing water potential, is acceptable.

(f) The student should plot a graph with the sucrose concentration on the x-axis and the percentage change in mass on the y-axis. She should then find the point where the curve crosses the x-axis and read off the sucrose concentration. This concentration is the water potential of the potato tissue.

ⓔ **2/3 marks awarded** The description of how to plot and use the graph is clear, but the sucrose concentration of the tissue is not the same as its water potential. The water potential could be found using a table or graph showing the relationship between sucrose concentration and water potential. The point where the curve crosses the x-axis is known as the x-axis **intercept**.

Student B

(a) So that the cells are genetically identical.

ⓔ **1/1 mark awarded** This is a different approach from student A's, but is an acceptable answer.

(b) She diluted 12 cm^3 of sucrose solution with distilled water.

ⓔ **0/1 mark awarded** Student B needed to state the volume of distilled water as well as the volume of sucrose solution.

(c) The 'skin' might have a different water potential from the other cells.

ⓔ **0/1 mark awarded** Student B would have gained this marking point if they had said that the skin *cells* might have a different water potential.

(d) So she can compare potato chips with different starting masses.

ⓔ **1/1 mark awarded** This answer is not as detailed as the one above, but is enough to be awarded the mark.

(e) Water has moved into the cells from a high concentration to a low concentration.

ⓔ **0/2 marks awarded** Although student B has correctly identified the direction of water movement, they have not used the terms *osmosis* or *water potential*, which are expected in an A-level answer.

(f) She should plot a graph of her results and find the concentration when there is no change in mass.

ⓔ **1/3 marks awarded** Student B has been awarded a mark for stating how to use the graph, but more detail should have been included about how to plot the graph.

Question 4 The effect of temperature on membrane permeability

A student carried out an investigation into the effect of temperature on the permeability of beetroot membranes. He:

1   cut ten discs of fresh beetroot and washed them thoroughly in distilled water.

2   collected five test tubes and measured $10\,cm^3$ of distilled water into each tube.

3   stood the test tubes in five separate water baths at 30, 40, 50, 60 and 70°C, and then left them for 5 minutes.

4   placed two beetroot discs in each tube and left them for 30 minutes, gently shaking the tubes every 5 minutes.

5   removed the beetroot discs after 30 minutes and poured each solution into a clean test tube.

6   used a colorimeter to measure the absorbance of each solution and then plotted a graph, as shown in Figure 3. The higher the absorbance, the more pigment has been released.

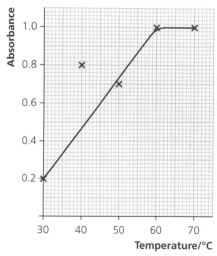

Figure 3 Graph plotted by a student to show the effect of temperature on the permeability of beetroot membranes

(a) Why did the student wash the beetroot discs in distilled water after cutting them? (1 mark)

e Cutting the discs will damage the membranes of some of the beetroot cells, so you need to think about the effect this might have if the student did not wash the discs.

(b) Why were the test tubes of distilled water left in the water baths for 5 minutes before adding the beetroot? (1 mark)

e The distilled water measured into each tube would have been at room temperature, so your answer needs to explain what will happen to the temperature of the distilled water after 5 minutes in the water bath.

(c) Why did the student shake the test tubes every 5 minutes? (1 mark)

e When you are answering this question, your main consideration should be a factor that will affect the diffusion of the pigment out of the cells.

(d) Why did the student use fresh beetroot and not cooked beetroot? (1 mark)

e Cooked beetroot will have been exposed to high temperatures, so you need to consider the effect that this would have had on both the cell-surface membrane and the tonoplast.

(e) The student thinks that one of his results is anomalous. Describe what he should do. (2 marks)

e You may have been taught at GCSE to ignore anomalous results, but they should not be ignored. Think about the steps you would take if one of your results was anomalous.

(f) Describe the effect of increasing temperature on the permeability of the beetroot cell membranes. (2 marks)

e This is a 'describe' question, so you need to state the overall shape of the curve *and* use figures from the graph.

(g) Explain the shape of the curve between 60°C and 70°C. (1 mark)

e The curve reaches a plateau, or levels off, between these two temperatures. Your answer should give reasons why there is no further increase in absorbance by referring to the concentration of the pigment, or the damage to the membrane.

Student A

(a) This is so that any pigment in the distilled water has been released from the cells due to the effect of temperature on membrane permeability.

e **1/1 mark awarded** Correct — the cutting will have released pigment from damaged cells.

(b) This will allow the water to equilibrate.

e **1/1 mark awarded** This is a good term to use.

(c) This is to maintain the diffusion gradient for the pigment to pass out of the cells and into the distilled water.

ⓔ **1/1 mark awarded** This is a good, clear answer.

(d) Cooking would involve the beetroot being heated to high temperatures that would denature the proteins in the plasma membranes and allow the pigment to leak out.

ⓔ **1/1 mark awarded** This is a clear answer, which explains how cooking would affect the membrane proteins.

(e) He should repeat the experiment in exactly the same way.

ⓔ **1/2 marks awarded** To gain the second mark, student A needed to state that the repeat is to check if the result is similar.

(f) There is a steady increase in permeability from 30°C to 60°C, then the curve reaches a plateau at an absorbance of 1.0.

ⓔ **2/2 marks awarded** The overall shape of the curve has been clearly described and suitable values have been used to support this description.

(g) There is no further diffusion of pigment out of the cells because the concentration of pigment outside the cells is equal to the concentration inside the cells.

ⓔ **1/1 mark awarded** This is a good answer, which gives a clear explanation in terms of diffusion and concentration.

Student B

(a) This will remove any pigment from the surface of the discs.

ⓔ **1/1 mark awarded** This is enough for the mark.

(b) So the water in each tube warms up to the required temperature.

ⓔ **1/1 mark awarded** Although the term 'equilibrate' has not been used, student B is still awarded the mark for this explanation.

(c) Shaking the tube will stop the discs from sticking together, which would reduce the surface area for diffusion.

ⓔ **1/1 mark awarded** Although student B has not referred to the diffusion gradient, this is still a relevant comment that is worthy of the mark.

(d) Cooking beetroot would damage the membranes and the pigment would already have been released.

ⓔ **1/1 mark awarded** Although student B has not referred to denaturing of the proteins, they have been awarded a mark for the effect on the membrane and a mark for stating how this would affect the pigment.

(e) He should circle this result and ignore it when he is calculating his mean.

ⓔ **0/2 marks awarded** Anomalous results happen and should not be ignored.

(f) The permeability increases and then levels off

ⓔ **1/2 marks awarded** Student B needed to include figures from the graph.

(g) All of the pigment has diffused out of the cells.

ⓔ **0/1 mark awarded** This answer is incorrect, but student B could have written that the pigment was diffusing in and out of the cells at the same rate.

Question 5 Estimating the number of stomata

A student observed the structures involved in gas exchange in a leaf. He removed a small piece of the lower epidermis and viewed it using an optical microscope. He counted the number of stomata that he could see in the field of view and used this to estimate the total number of stomata present in the lower epidermis of the leaf. Figure 4 shows the lower epidermis as viewed by the student.

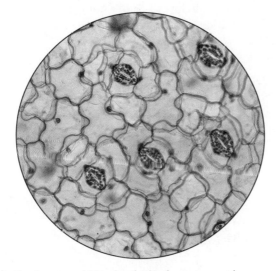

Figure 4 Stomata in the lower epidermis of a leaf, as seen using an optical microscope

(a) Describe how the student could prepare a piece of the lower epidermis to observe using an optical microscope. (3 marks)

ⓔ This question requires you to describe the method you would use to prepare a slide. Your answer should be logical and include the terms 'slide' and 'coverslip'.

(b) The student used an optical microscope with an eyepiece lens magnification of ×10 and objective lenses of ×4, ×10 and ×40. Calculate the maximum magnification that the student used. (1 mark)

ⓔ Magnification is calculated using both the eyepiece and the objective lens.

(c) Describe how the student could have used a clear ruler with mm divisions to find the area of the field of view using the low-power objective lens. (2 marks)

ⓔ The field of view refers to the circle that you can see when you look down an optical microscope. One of the maths skills that you may be assessed on is your ability to calculate the area of a circle. You will need to state the measurement you will take and then how you will use this to calculate area.

(d) The student found that there were five stomata per mm². Describe how he could have estimated the total number of stomata in the lower epidermis of one leaf. (3 marks)

ⓔ To answer this question, you firstly need to describe how you would find the area of the leaf. You then need to use the number of stomata given to describe how you would estimate the total number of stomata.

Student A

(a) He should take a clean glass slide and add one drop of water to the centre of the slide. He then needs to peel off the lower epidermis, which is thin as it is a single layer of cells, and lay it on the slide. Finally, he should cover the epidermis with a coverslip.

ⓔ **2/3 marks awarded** This is a well-structured answer, but needs more detail in the final sentence to minimise the risk of trapping air bubbles — for example, position the coverslip on the slide at an angle of 45° to the specimen and then gently lower it onto the slide using a mounted needle.

(b) total magnification = eyepiece lens × objective lens = 10 × 40 = ×400

ⓔ **1/1 mark awarded** It is always a good idea to show your working.

(c) He could put the ruler on the stage and focus on the scale, then measure the diameter of the field of view. He can then use πr^2 to calculate the area of the field of view.

ⓔ **2/2 marks awarded** This is a good answer that describes both steps clearly.

(d) He could use a piece of squared paper to draw round the leaf. Then he could count the number of squares to find the total area of the leaf in mm². He knows that there are five stomata in each mm², so he can multiply the area of the leaf by 5.

ⓔ **3/3 marks awarded** This is a concise, clear answer.

Student B

(a) The leaf should be put on a glass slide with a drop of water, then a coverslip put over it.

ⓔ **1/3 marks awarded** Student B understands the overall process, but this answer omits two important steps: the plant tissue must be thin, so just a piece of lower epidermis should be used and not the whole leaf; the method must avoid trapping air bubbles.

(b) ×400

ⓔ **1/1 mark awarded** The mark is awarded for the correct answer.

(c) He could look at the ruler down the microscope and measure the width of the field of view and then use this to find its area.

ⓔ **1/2 marks awarded** Student B needed to describe how to find the area by stating the equation.

(d) Find the area of the leaf in mm², then times by 5 to get total number of stomata.

ⓔ **2/3 marks awarded** Student B needed to describe *how* the area of the leaf could be found.

Question 6 The effect of antibiotics on bacterial growth

A student investigated the effect of antibiotics on the growth of bacteria. She used aseptic techniques to:

1 transfer 1 cm³ of liquid bacterial culture to an agar plate

2 spread the bacterial culture to cover the plate

3 place four paper discs soaked in antibiotic on the plate

The student incubated the agar plate at 25°C for 48 hours. Figure 5 shows the appearance of the plate following incubation.

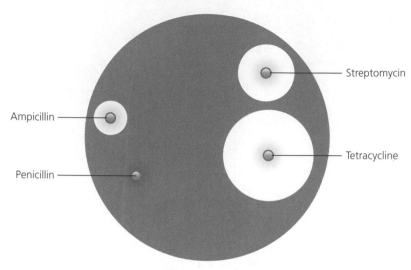

Figure 5 A diagram to show the appearance of an agar plate with antibiotic discs following incubation

(a) Describe *two* aseptic techniques that the student could have used. (2 marks)

e You are expected to have used aseptic techniques to minimise contamination and should be able to describe some of the methods that you used.

(b) Give *one* reason why the student used aseptic technique. (1 mark)

e Be specific when answering this question and avoid vague references to bacteria escaping.

(c) The student incubated the agar plates at 25°C. Suggest *one* reason why she did not incubate them at a higher temperature. (1 mark)

e A higher temperature would be closer to human body temperature. Your answer needs to give a possible reason for the student using a lower temperature.

The student measured the diameter of the clear zone around each antibiotic and recorded the results in Table 2.

Table 2 The diameter of clear zones around antibiotic discs

Antibiotic	Mean diameter of the clear zone/mm
Ampicillin	9
Streptomycin	15
Tetracycline	24
Penicillin	0

(d) Calculate the area of the clear zone around tetracycline. Use $\pi = 3.14$. (1 mark)

e Calculating the area of a circle is one of the maths skills named in the specification. You are expected to know the formula.

(e) The student concluded that penicillin was not effective against this bacterium. Explain why she reached this conclusion. (1 mark)

e Your answer should not just be a statement that there is no clear zone; it should include an explanation *why*.

Student A

(a) Use a sterile pipette to transfer the bacterial culture to the agar plate. Flame the neck of the bottle containing the bacterial culture.

e **2/2 marks awarded** These are two clearly described techniques that show student A's familiarity with the practical.

(b) This technique minimises the risk of bacteria from the air contaminating the agar plate.

e **1/1 mark awarded** This is a well-expressed answer, with good use of the term 'contaminating'.

(c) A higher temperature could be closer to the optimum temperature for bacteria that cause human disease. This would be dangerous for the student.

e **1/1 mark awarded** This is exactly why the incubation temperature should not exceed 25°C.

(d) Diameter of clear zone for tetracycline = 24 mm, so radius = 12 mm.
area of clear zone = πr^2 = 3.14 × 12^2 = 452.16 mm^2

e **1/1 mark awarded** Although the mark would be awarded for giving the correct answer, it is good practice to show your working in this way.

(e) There is no clear zone around the penicillin, so this antibiotic did not inhibit the growth of the bacteria.

e **1/1 mark awarded** Another well-written answer, with good use of the term 'inhibit'.

Student B

(a) Work next to a lit Bunsen burner. Wash hands thoroughly before and after the practical.

e **1/2 marks awarded** Student B's first point is specific to microbial aseptic technique, but hand washing is an example of good practice that relates to all practical activities. A better answer could refer to wiping down the bench with disinfectant before and after the practical.

(b) These techniques protect the student from harmful bacteria.

🔴 **0/1 mark awarded** This answer is too vague to gain the mark. A more specific answer might say 'these techniques prevent the release of harmful bacteria into the air'.

(c) This is the optimum temperature for bacteria, so they will grow better.

🔴 **0/1 mark awarded** Although this may be the optimum temperature for some bacteria, this answer does not reflect the understanding expected at A-level.

(d) 75.36 mm^2

🔴 **0/1 mark awarded** This is an incorrect answer because student B did not use the correct formula. They have used πd, which is the formula for the circumference of a circle, and have not shown any working.

(e) The bacteria could grow next to this antibiotic, showing that penicillin did not kill the bacteria or slow their growth.

🔴 **1/1 mark awarded** This is a simply expressed answer, but it shows understanding of the practical results and gains the mark.

Question 7 Using chromatography to investigate leaf pigments

A student investigated the pigments present in two different leaves, A and B. The steps he followed were:

1 Crush each leaf with solvent to extract the pigments.

2 Label two strips of filter paper A and B.

3 Draw a pencil line 1 cm from the bottom of each strip of paper.

4 Use a fine glass tube to put a spot of extract from leaf A on the pencil line on paper A.

5 Put a spot of extract from leaf B on the pencil line on paper B.

6 Stand each strip of filter paper in a boiling tube of solvent; check that the solvent is not above the pencil line.

7 Leave the strips of filter paper in the boiling tubes for 30 minutes without moving them.

8 Remove each strip of filter paper and draw a pencil line to show how far the solvent has travelled.

Figure 6 shows the appearance of one of the completed chromatograms.

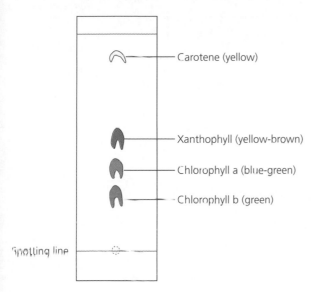

Spotting line

Figure 6 Plant pigments separated on a paper chromatogram

(a) Explain why the student used pencil to draw the spotting line and not pen. (1 mark)

ⓔ When answering this question, remember that the ink used in pens might be soluble in the solvent you are using.

(b) Explain why the student checked that the solvent in the boiling tube was not above the pencil line. (1 mark)

ⓔ The pigment spot is on the pencil line, so you need to describe what would happen to the pigment if the solvent were above it.

(c) Suggest why the student did not move the boiling tubes once the filter paper was inside. (1 mark)

ⓔ Moving the boiling tubes will move the solvent in the bottom of the tubes. Your answer needs to describe how the solution not being level might affect the movement of either the solvent or the pigments.

(d) Explain why the student drew a line to show how far the solvent travelled as soon as he removed it from the tube. (1 mark)

ⓔ You need to know how far the solvent has travelled to calculate the Rf value. You need to think about what will happen to the solvent on the filter paper once it has been removed from the tube.

(e) The student wants to calculate the Rf value of a pigment so he can compare his results with those of another student. Describe how he would calculate the Rf value of a pigment. (2 marks)

ⓔ Look at the number of marks available for this question. You need to state the measurements that you would need to make and how you would use these measurements.

Questions & Answers

Student A

(a) Pencil would not dissolve in the solvent.

🅔 **1/1 mark awarded** This is a clear and concise answer that gains the mark.

(b) The pigments found from the leaves are soluble in the solvent and would be removed from the paper if the solvent were above the pencil line.

🅔 **1/1 mark awarded** This answer makes good use of the terms 'soluble' and 'solvent'.

(c) Moving the tubes would move the solvent in the bottom of the tube and might cause it to travel up the paper at an angle instead of running straight.

🅔 **1/1 mark awarded** This is a good point — the solvent travelling at an angle might affect the movement of the pigment and the calculation of Rf values.

(d) The solvent would evaporate and you would not be able to see the solvent front.

🅔 **1/1 mark awarded** Good use of the term 'solvent front' — this term was not used in the question to describe the distance travelled by the solvent.

(e) He should measure the distance from the origin to the centre of the pigment spot and measure the distance from the origin to the solvent front. The Rf value is calculated using distance travelled by pigment divided by distance travelled by solvent.

🅔 **2/2 marks awarded** Measuring to the centre of the pigment spot is an excellent way of standardising the procedure, but the answer would still have gained full marks without this point. Good use of terminology again: for example, 'origin' was not used in the question to describe the pencil line.

Student B

(a) The pen might travel up the filter paper.

🅔 **0/1 mark awarded** Student B has an idea of what might happen, but this is poorly expressed and no mark is awarded. A better answer might say 'the ink from the pen might run with the solvent'.

(b) If the solvent was above the pencil line it might wash the pigments off.

e 1/1 mark awarded This is a basic answer, but is worth 1 mark despite the simple terms used.

(c) The pigments might run off the side of the paper.

e 1/1 mark awarded This is another basic answer that just gains the mark.

(d) The paper will dry quickly and you will not be able to see how far the solvent travelled.

e 1/1 mark awarded Student B has not used the terms 'evaporate' or 'solvent front', but has included the key idea and gained the mark.

(e) The distance travelled by the solvent is divided by the distance travelled by the pigment.

e 1/2 marks awarded A mark is awarded here for the implication that these distances have to be measured, but student B has the measurements the wrong way round.

Question 8 Investigating photosynthesis using a chloroplast suspension

A student carried out an investigation using a chloroplast suspension and a blue dye called DCPIP. When DCPIP is reduced, it changes colour from blue to colourless. The student made a suspension of chloroplasts by blending leaves with ice-cold isolation medium and then filtering. She set up three boiling tubes, as shown in Table 3, covered tube 2 with foil, and then stood all three tubes next to a bright light for 2 hours. Table 3 shows the contents of each tube and its appearance at the beginning and end of the investigation.

Table 3 The effect of a chloroplast suspension on the colour of DCPIP

Tube number	Contents of the tube	Appearance at the start of the investigation	Appearance after 2 hours
1	Chloroplast suspension Buffer at pH 7 DCPIP	Blue	Green
2	Chloroplast suspension Buffer at pH 7 DCPIP	Blue	Blue
3	Boiled chloroplast suspension Buffer at pH 7 DCPIP	Blue	Blue

Questions & Answers

(a) Explain the colour change in tube 1. (2 marks)

ⓔ The DCPIP has been decolourised, meaning that the DCPIP has been reduced. You need to use your knowledge of chlorophyll and the electron transfer chain to explain how DCPIP was reduced.

(b) The student covered tube 2 with foil. Explain why the DCPIP in this tube remained blue. (2 marks)

ⓔ DCPIP will decolourise when it accepts electrons. Your answer needs to demonstrate your understanding of the transfer of electrons during photosynthesis.

(c) Explain why the student used isolation medium that was ice-cold. (2 marks)

ⓔ Blending the leaves will break open the leaf cells, releasing the cell contents, including enzymes. Your answer needs to make the link between enzyme activity and the isolation medium being ice-cold.

(d) The isolation medium contained sucrose. Explain why sucrose was included in the isolation medium. (2 marks)

ⓔ Your answer needs you to explain what would happen if the chloroplasts were in water rather than in a sucrose solution. How would chloroplasts be affected by water entering or leaving them?

(e) Explain why the DCPIP in tube 3 did not decolourise. (2 marks)

ⓔ The chloroplast suspension in tube 3 had been boiled. Your answer needs to include the effect of boiling on proteins and how this would affect photosynthesis.

Student A

(a) The DCPIP has decolourised so has been reduced. This means that DCPIP has accepted electrons released from chlorophyll that have been transferred along the electron transfer chain.

ⓔ **2/2 marks awarded** Student A has gained marks here for their understanding of the decolourisation of DCPIP and for explaining the movement of electrons.

(b) The foil prevented light from reaching the chlorophyll, so photoionisation did not take place and no electrons were released from the chlorophyll molecule. This meant that no electrons were available to reduce DCPIP, so it stayed blue.

ⓔ **2/2 marks awarded** This is a detailed, well-written answer that demonstrates a good understanding of the role of light in photosynthesis.

(c) Blending the leaves releases enzymes from the cell that could damage the chloroplasts. An ice-cold buffer will reduce the rate of enzyme activity and prevent chloroplast damage.

e **2/2 marks awarded** Releasing enzymes such as lipases and proteases, which are normally compartmentalised within lysosomes, could lead to the breakdown of the chloroplast membranes. Reducing the activity of these enzymes is essential for the chloroplasts to continue functioning.

(d) The presence of sucrose reduced the water potential of the isolation medium. This meant that there was no net movement of water into the chloroplasts. If water entered the chloroplasts by osmosis, they would burst.

e **2/2 marks awarded** This is an excellent answer that shows clear understanding of the term 'water potential' and relates this to an effect on the chloroplasts.

(e) Boiling would denature the proteins in the chloroplasts, including those that make up the electron transfer chain. Electrons would not pass along the electron transfer chain and so cannot be accepted by DCPIP.

e **2/2 marks awarded** The first mark is awarded for explaining that the proteins in the electron transfer chain are denatured, and the second mark for explaining why this would prevent the reduction of DCPIP.

Student B

(a) The DCPIP has lost its blue colour. This means that it has been reduced, so has gained electrons.

e **0/2 marks awarded** Stating that DCPIP changes colour when it is reduced is just repeating the stem of the question, so no mark is awarded for this. Student B knows that reduction is gain of electrons, but this is not worthy of a mark at A-level. Some reference to chlorophyll as the source of the electrons should have been made.

(b) The foil blocks out light, so no photosynthesis can occur.

e **0/2 marks awarded** Although student B makes the link between blocking out light and photosynthesis, this answer lacks the necessary details and terminology that student A includes in their answer.

(c) This will reduce the kinetic energy of the enzymes, so chemical reactions will happen more slowly.

e **1/2 marks awarded** A mark is awarded for recognising that the low temperature will reduce the rate of enzyme activity, but this needs to be linked to an effect on the chloroplasts.

(d) Having sucrose in the isolation medium means that no osmosis will take place, so no water will move into or out of the cell. If water moved into the cells by osmosis, the cells might burst.

ⓔ **0/2 marks awarded** Although student B recognises that the presence of sucrose has an effect on osmosis, there is no reference to water potential and saying that there is 'no osmosis' is poor phrasing. This answer does not gain the second mark because student B refers to cells bursting rather than describing the effect on chloroplasts; the leaf cells have already been ruptured by blending, and plant cells would not burst anyway due to their cell walls.

(e) The chloroplast suspension was boiled in tube 3 and proteins are denatured at high temperatures. The light-dependent reaction of photosynthesis would not take place.

ⓔ **1/2 marks awarded** This answer is awarded 1 mark for correctly describing the effect of temperature on proteins, but student B has not fully answered the question because they have not explained why this would prevent the reduction of DCPIP.

Question 9 Using a respirometer to investigate oxygen uptake

A student put 2 g of yeast into a glucose solution. He used the apparatus shown in Figure 7 to investigate the rate of oxygen uptake by the culture of yeast at 30°C.

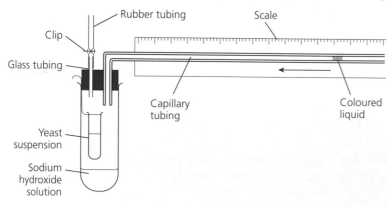

Figure 7 A simple respirometer

(a) The drop of coloured liquid moves through the capillary tubing towards the yeast suspension (as shown by the arrow in Figure 7). Explain why the liquid moves in the direction shown.

(3 marks)

ⓔ This question is about pressure differences. The yeast is respiring aerobically and so is taking in oxygen and releasing carbon dioxide. The carbon dioxide is absorbed by the sodium hydroxide. How would this affect the pressure inside the tube compared with outside the tube?

The student recorded the volume at the start of his investigation and took a reading from the scale every 10 minutes. His results are shown in Table 4.

Table 4 Volume readings from a respirometer taken at 10-minute intervals

Time/min	Volume reading on the scale/cm^3
0	3.2
10	3.8
20	4.5
30	5.0
40	5.4
50	5.6
60	5.7

(b) Plot a graph of these results. (A grid will be provided in the exam paper.) (5 marks)

ⓔ Marks may be available for having the axes the correct way round, choosing a suitable scale, labelling the axes, accurately plotting points and drawing a suitable line.

(c) Calculate the rate of oxygen use per gram of yeast during this investigation. Include a suitable unit. (2 marks)

ⓔ You need to use two numbers to calculate the rate: the volume of oxygen taken up in 1 hour and the mass of yeast.

(d) The student repeated the investigation at 20°C and found that the volume of oxygen taken up by the yeast was lower. Explain why. (1 mark)

ⓔ This question requires you to apply your knowledge and make the link between temperature and respiration.

Student A

(a) The yeast respires aerobically, taking in oxygen and releasing carbon dioxide. Carbon dioxide is absorbed by potassium hydroxide decreasing the pressure inside the tube so that it is less than atmospheric pressure and the liquid moves to the left (as shown by the arrow).

ⓔ 3/3 marks awarded The marks are awarded for stating that oxygen is taken in, and that carbon dioxide is absorbed, and for explaining how this affects pressure.

(b)

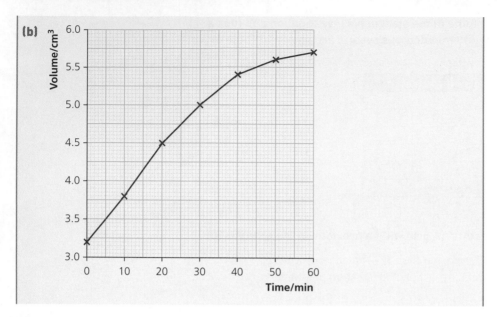

ⓔ 5/5 marks awarded The axes are the correct way round, with the IV on the x-axis; a suitable scale has been chosen, making good use of the grid; axes are fully labelled and units included; points are plotted clearly and correctly; a suitable line has been drawn.

(c) rate of oxygen use = 5.7 – 3.2 = 2.5 cm³ in 1 hour
There are 2 g of yeast, so oxygen use = 1.25 cm³ g⁻¹ hour⁻¹.

ⓔ 2/2 marks awarded Working has been clearly shown and a suitable unit chosen. The rate could also be given per minute: 1.25/60 = 0.02 cm³ min⁻¹ g⁻¹.

(d) The rate of respiration is higher at 30°C because this is closer to the optimum temperature for the enzymes involved in respiration.

ⓔ 1/1 mark awarded This is an excellent answer, with good use of scientific terminology.

Student B

(a) The liquid moves because the carbon dioxide produced by respiration is absorbed by the potassium hydroxide, creating a vacuum that sucks the liquid along the tube.

ⓔ 1/3 marks awarded Student B has been awarded a mark for stating that the carbon dioxide has been absorbed, but there is no reference to oxygen uptake or the effect on pressure. The term 'vacuum' is incorrect and 'sucks the liquid along' is a poor expression.

(b)

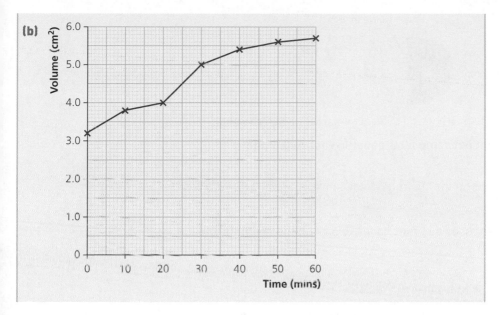

2/5 marks awarded The axes are the correct way round and a suitable scale has been chosen, but the y-axis scale means that there is a lot of unused space. Marks have been lost because the point at 20 minutes has been plotted incorrectly and the wrong unit has been used on the y-axis (cm^2 instead of cm^3). The units should have a solidus before them rather than being written in brackets, but you would not be penalised marks for writing them in this format.

(c) 2g of yeast takes up $2.5 \, cm^3$ of oxygen in 60 minutes, so 1g of yeast takes up $0.02 \, cm^3$ of oxygen in 1 minute.

ⓔ **1/2 marks awarded** Student B shows a good understanding of the data and this is a clearly presented calculation, but the unit for rate has not been written correctly as $cm^3 \, min^{-1} \, g^{-1}$.

(d) Less oxygen is taken up because the yeast is respiring less.

ⓔ **0/1 mark awarded** This is insufficient for a mark because there is no explanation of *why* the rate of respiration is lower.

Question 10 The effect of light on maggot behaviour

A student investigated the effect of light on the behaviour of maggots using a choice chamber. She put damp filter paper in the bottom of the choice chamber, then covered half of the chamber with light-proof black cloth and left the other half uncovered (Figure 8). Using a small plastic spoon, she added 12 maggots to the chamber and left them for 10 minutes. She recorded the number of maggots in each half of the choice chamber every minute for 10 minutes.

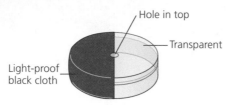

Figure 8 A diagram to show how the student set up the choice chamber

(a) Suggest why the student put damp filter paper in each half of the choice chamber. (1 mark)

ⓔ Your answer should focus on the environmental conditions, while writing in an appropriate scientific style.

(b) Suggest why the student used a plastic spoon to pick up the maggots and not a pair of forceps. (1 mark)

ⓔ The ethical use of organisms is one of the skills you are expected to have developed, so your answer should consider the effect on the animals.

Table 5 Mean number of maggots present in each half of the choice chamber

Environmental condition	Mean number of maggots present per minute
Dark	10
Light	2

(c) Table 5 shows that more maggots were recorded in dark conditions than light. Explain why this behaviour would be an advantage to maggots. (1 mark)

ⓔ Your answer should be specific and should indicate how this response to light would increase the organism's chance of survival.

(d) The student wanted to determine whether there was a significant difference between her observed and expected results.

   (i) State a null hypothesis for the student's investigation. (1 mark)

ⓔ Remember that a null hypothesis is a testable statement that there would be no difference.

   (ii) Use the null hypothesis to state the expected number of maggots in each area of the choice chamber after 10 minutes. (1 mark)

ⓔ Refer to your null hypothesis from part (i) and the total number of maggots given in the question.

   (iii) State the statistical test that the student should use to analyse her results. Give the reason for your choice of test. (2 marks)

ⓔ The first mark is for simply stating the name of the statistical test. The second mark is for giving your reason. The maths skills section for required practical 10 gives information about the three statistical tests you need to know and when they should be used.

Student A

(a) Damp conditions provide a more favourable environment for maggots as this prevents them from drying out.

e **1/1 mark awarded** This is a good, concise answer.

(b) She might squeeze a pair of forceps too hard and harm the maggots.

e **1/1 mark awarded** This is a simple example of the ethical use of animals.

(c) If the maggots move towards dark regions, they are less likely to be seen by predators and this increases their chance of survival.

e **1/1 mark awarded** It also encourages them to burrow into their food and avoid drying out.

(d) (i) There is no difference in the number of maggots found in light and dark conditions.

e **1/1 mark awarded** This is a concise null hypothesis.

(ii) There would be six maggots in each half.

e **1/1 mark awarded** In other words, the 12 maggots divided equally between each half.

(iii) She needs to use the chi-squared test because the data are categoric and she is comparing the frequencies.

e **2/2 marks awarded** This is a good explanation for the choice of statistical test. Student A has also correctly used the word 'data' as a plural (data *are* categoric, not data *is* categoric).

Student B

(a) This is because maggots like damp conditions.

e **0/1 mark awarded** Avoid attributing human feelings and emotions to simple animals. The maggots do not 'like' damp conditions, but it is their preferred environment.

(b) The maggots might get stressed if they were picked up using forceps.

e 0/1 mark awarded Describing maggots as 'stressed' is poor biology and not worthy of a mark at A-level.

(c) The maggots are trying to avoid predators by hiding in the dark.

e 0/1 mark awarded This is another example of poor biology, with the suggestion that maggots are intentionally moving to the darker environment to avoid predation.

(d) (i) There is no significant difference between the results.

e 0/1 mark awarded This is too vague and makes no reference to the different conditions.

(ii) I would expect most of the maggots to be in the dark.

e 0/1 mark awarded The question relates to a statistical test, so the term 'expected' should be used in that context rather than interpreted as what you expect to happen.

(iii) The stats test she should use is the chi-squared test because she is comparing observed and expected results.

e 1/2 marks awarded Only the mark for the choice of test has been awarded because the reason for the choice is too vague. Student B should have referred to the type of data — categoric.

Question 11 Using a calibration curve to find glucose concentration

A student was given glucose solutions of concentrations 2, 4, 6, 8 and 10%. The method followed by the student is given below:

1   Add dilute sulfuric acid and a solution of potassium manganate(VII) to each glucose concentration.

2   Time how long each concentration takes to change colour from pink to colourless.

3   Repeat steps 1 and 2 for each concentration.

4   Plot a graph of the results (Figure 9).

5   Use this graph to estimate the concentration of an unknown glucose solution.

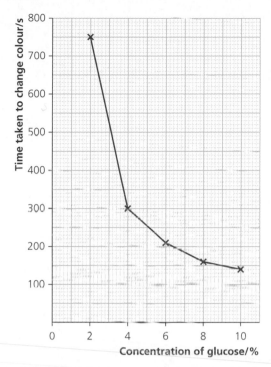

Figure 9 Graph of time taken to change colour against glucose concentration

(a) The student had to decide when the pink colour had completely disappeared. Suggest how he ensured that his decision was consistent. (1 mark)

ⓔ There are different ways of approaching this question, but your answer could include a colour comparison.

(b) Explain why the student repeated steps 1 and 2 for each concentration. (1 mark)

ⓔ This question asks you to *explain*, so you need to make sure that your answer includes enough detail — 'to calculate the mean' would be insufficient.

(c) The student estimated the concentration of the unknown solution to be 5%. Explain why this was only an estimate. (1 mark)

ⓔ The points on the graph have been joined, because the student could not be certain of the intermediate values. Your answer should refer to concentrations that have not been tested.

(d) Explain how the student could modify his method to find a more accurate concentration of the unknown solution. (1 mark)

ⓔ The method should be modified to include more concentrations of glucose, but your answer should specify the range of concentrations.

Student A

(a) He could use a solution that had already completely decolourised, and then compare all of the other solutions with this one.

ⓔ 1/1 mark awarded This technique is called using a colour standard and would allow a consistent end-point to be determined.

(b) This was so he could show that his measurements were repeatable and could identify any anomalous results.

ⓔ 1/1 mark awarded The word 'repeatable' is used well here because the student repeated the experiment using the same equipment and method. Measurements would be **reproducible** if another person carried out the investigation and obtained the same results.

(c) It is only an estimate because the student did not test a 5% glucose concentration.

ⓔ 1/1 mark awarded This concise answer gains the mark.

(d) He should repeat the experiment, but using more concentrations between 4 and 6%, for example 4.5, 5.0 and 5.5%.

ⓔ 1/1 mark awarded This detailed answer would still gain the mark for just stating 'values between 4 and 6%' without specifying the concentrations.

Student B

(a) He could put a piece of white card behind each solution to check that there is no pink colour left.

ⓔ 1/1 mark awarded This answer is more simply expressed than student A's, but is still worthy of a mark.

(b) Having repeats allows you to identify anomalous results. These can be ignored when you are calculating the mean.

ⓔ 0/1 mark awarded Anomalous results should not be ignored — possible reasons for the anomaly should be investigated.

(c) The student did not test the intermediate values between 4 and 6%.

ⓔ 1/1 mark awarded This answer is phrased slightly differently from student A's response, but still gains the mark.

(d) He should test more concentrations around 5%.

ⓔ 1/1 mark awarded Although student B has not suggested specific concentrations, 'around 5%' gives enough information to gain the mark.

Question 12 The effect of distance from a tree on leaf size

A student investigated the effect of distance from a tree on leaf shape and size. She measured the length of dandelion leaves at different distances from a tree and counted the number of lobes present on each leaf (Figure 10).

(a)

(b)

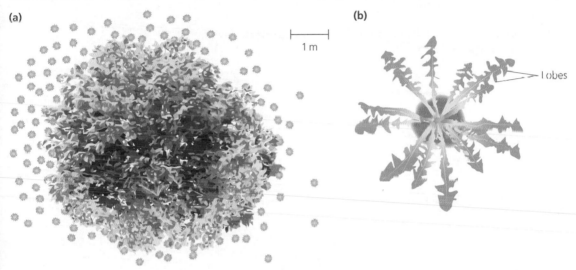

1 m

lobes

Figure 10 (a) A tree surrounded by dandelion plants, as viewed from above.
(b) A dandelion plant as viewed from above, showing leaf lobes

(a) (i) Name an environmental variable that could affect leaf length. (1 mark)

ⓔ You could name any biotic or abiotic factor that could realistically have an effect on the size of leaves.

(ii) Explain how the variable named in (i) could affect leaf length. (1 mark)

ⓔ Make sure that your answer is linked to the variable you gave in (i) and describes its effect on a biological process related to leaf size, such as photosynthesis or respiration.

(b) The student measured the length of the dandelion leaves. Give *one* reason why leaf length may not be the best way to measure leaf size. (1 mark)

ⓔ You just need to give the name of one other measure of leaf size that could vary.

(c) Suggest *one* reason why the student measured the leaves while they were still attached to the plant. (1 marks)

ⓔ Your answer could refer to an ethical reason for not removing the leaves, or to the effect removing the leaves would have on the measurements.

(d) The student plotted a graph of distance from the base of the tree in metres against the mean dandelion leaf length in mm.

Describe how the student could have used a quadrat and tape measure to collect the data for this graph.

(2 marks)

ⓔ To plot the graph, the student would have to measure the length of leaves at specific distances from the tree, so your answer should describe the use of a belt transect. This is *not* random sampling.

The student plotted a graph of leaf length against the number of lobes per leaf and concluded that there was a positive correlation.

(e) (i) Name the type of graph that the student should have plotted. Give a reason for your answer.

(1 mark)

ⓔ Notice that you need to name the type of graph *and* explain your choice of graph. Make sure that you give a specific reason related to the type of data collected.

(ii) What does *positive correlation* mean?

(1 mark)

ⓔ This is a straightforward recall question, assessing your understanding of this term.

Student A

(a) (i) light intensity

ⓔ 1/1 mark awarded Student A has correctly identified a factor that could affect leaf size.

(ii) Light intensity would affect the rate of photosynthesis, which may affect leaf length.

ⓔ 1/1 mark awarded A clear link has been made to photosynthesis.

(b) Leaves might be different widths as well as different lengths.

ⓔ 1/1 mark awarded Student A recognises that length is not the only measurement of leaf size.

(c) Picking the leaves might damage the plant. The leaves might dry out and shrink once they have been picked, which would reduce their size.

ⓔ 1/1 mark awarded Student A has included two acceptable points here, but only 1 mark is available.

(d) To collect these data, she would have laid the tape measure out perpendicular to the tree and positioned the quadrat at regular intervals, about every 2 m. She would then measure the length of the leaves and find the mean leaf length.

ⓔ 2/2 marks awarded The technique is clearly described, with the correct uses of both the tape measure and the quadrat.

(e) (i) She should draw a scatter graph because she is looking for a relationship between two continuous variables.

ⓔ 1/1 mark awarded This is a good answer that names the type of graph and gives a clear reason for the choice.

(ii) Positive correlation means that as one variable increases, so does the other variable.

ⓔ 1/1 mark awarded Student A's answer gains the mark for a general explanation of the term

Student B

(a) (i) Temperature

ⓔ 1/1 mark awarded This is another abiotic factor that would affect leaf size. Examples of biotic factors could be predation or disease.

(ii) Respiration involves a series of enzyme-controlled reactions, so it would be affected by temperature. Less respiration might lead to less leaf growth.

ⓔ 1/1 mark awarded The mark has been awarded for linking temperature to respiration, but it would be better to say 'a reduced rate of respiration' than 'less respiration'.

(b) Some leaves might be wider or thicker.

ⓔ 1/1 mark awarded This is a good point, especially the reference to variation in leaf thickness.

(c) Picking the leaves harms the environment, and she might mix up the leaves and forget where she picked them.

ⓔ 0/1 mark awarded Neither of these points is worthy of a mark. The term 'environment' is too vague, whereas referring to damage to the *habitat* or *ecosystem* would gain a mark. References to human error that suggest poor experimental technique would not be credited at A-level.

(d) She would have used the tape measure to map out a grid and used the quadrat to take random samples.

ⓔ 0/2 marks awarded Student B has described a random sampling technique that is not appropriate for this investigation.

(e) (i) The student should have drawn a scattergram because she is looking for a relationship between pairs of measurements.

ⓔ 1/1 mark awarded Scattergram, scatter graph and scatter diagram are all terms that could be used to describe this type of graph. Although student B has not included the term 'continuous' when describing the variables, they still gain a mark for stating that the measurements are in pairs.

(ii) This means that as the length of the leaf increases, the number of lobes also increases.

ⓔ 1/1 mark awarded Student B has linked the answer to the example given and gains the mark.

Index

membrane permeability   29–33, 76–79

meristematic cells   20

methylene blue dye as electron acceptor   46–47

microscope use   21–22

middle lamella   **20**

mitosis   19, 22, 23

mitotic index, calculating   19, 22

N

null hypothesis   7, **52**

nutrient broth   37

O

optical microscopes, using   21–22

osmosis   24–25, 73–76

oxidation   42

P

paired *t*-test   55

percentage change   29

percentage error   12

pH effects

amylase activity   16

membrane permeability   29–30

starch hydrolysis   67–70

photoionisation   42

photolysis   42

photosynthesis   39–40, 42, 87–90

pigments in plant tissue

extracting using solvents   39–42, 84–87

membrane permeability   29–33, 76–79

precision   **11**

proportional dilution   26

protease   13

Q

quadrats   61, 62

quantitative results   30

quantitative technique, colorimeters   32

R

random errors   11

random sampling   61

range bars   60

reaction rates   13–19

recording observations   10

reduction   42

repeatable results   11–12

reproducible results   12

resolution   8

respiration rate, factors affecting   44–48

respirometers, using   46, 90–93

Rf values   41, 42

risk assessment   4, 9, 35, 51

root tip squash, preparing and observing   19–24, 70–72

S

safety issues   9, 34–35

sampling bias   61

sampling techniques   61–62

scattergrams   63–64

semi-quantitative technique, colour standards   31

serial dilution   26, 39

signs   **57**

slide preparation   20–21

solvents, chromatography   40–41, 84–87

Spearman's rank correlation test   55, 64–66

species distribution   60–63

staining   21

statistical tests   53–57

Student's *t*-test   54, 55–57

subjective results   **31**

sucrose concentration

conversions to water potential   27

effect on osmosis   73–77

surface areas, calculating   48

symptoms   **57**

systematic errors   10

systematic sampling   62

T

tables, drawing   10, 14–15, 27, 47

taxes   49

temperature and membrane permeability   29–33, 76–79

times, recording   44

t-tests   53–57

U

uncertainty   12, 14

representing   60

uncontrolled variables   11

unpaired *t*-test   55

V

validity   12

volume calculations   25, 48

W

water potential, plant tissue   24–28, 74–76

Z

zones of inhibition   38, 39